Praise for
SLOW MOTION

"Chilling. . . . There is a gritty honesty to her cautionary confession that will alert others to listen for and respond to wake-up calls of their own."

—*New York Times Book Review*

"Riveting. . . . A combination of breathtaking candor and bravado. . . . A smart, well-written take on the form."

—*San Francisco Chronicle*

"A candid memoir of a misspent life resurrected from the ruins of a devastating tragedy. . . . Shapiro does not sugarcoat her life; she writes with an eviscerating, raw honesty about the wrong turns and mistakes she made."

—*Boston Globe*

"Fabulous. . . . Brutal honesty and a well-honed sense of humor."

—*People*

"Stylishly written . . . splendid. . . . Shapiro's prose is seamless."

—*Newsweek*

"Absorbing, sweetly stinging. . . . Shapiro's book succeeds as a gracefully written story of reckoning inspired by tragedy and the long reach of familial roots."

—*Wall Street Journal*

"Cogent and unforgettable."

—*Houston Chronicle*

"Shapiro's writing has the spare elegance of a thin, gold bracelet—with all the timeless appeal and elegance and fine craft that implies. Her self-examination is untainted by self-pity. . . . A great piece of writing and an inspirational tale. . . . Slow Motion illuminates the rocky road to integrity in graceful but wrenching steps."
—*Salon*

"Notably free of self-pity and rigorous in her scrutiny, Shapiro doesn't spare her lover, her family, or herself, yet there's emotion on these pages that is rare among the recent spate of confessionals. The result is a story that isn't perversely voyeuristic, but simply rewarding."
—*Entertainment Weekly*

"Well crafted and full of juice."
—*Newsday*

"A page-turner."
—*Mademoiselle*

"An unnerving lost-girl memoir told with humor and aching bafflement."
—*Elle*

"Shapiro is an uncommonly fine writer who seems effortlessly to combine a lean prose style with acute observational detail. . . . Shapiro's book should be celebrated not as a morality tale, but as good literature."
—*The Guardian* (London)

"Remarkable and painful. . . . Shapiro lays bare her own young and fractured self with searing candor. The result is moving, comic, and bitterly self-aware."
—*The Times* (London)

SLOW
MOTION

SLOW MOTION

A Memoir of a Life
Rescued by Tragedy

DANI SHAPIRO

HARPER PERENNIAL

NEW YORK • LONDON • TORONTO • SYDNEY • NEW DELHI • AUCKLAND

HARPER ● PERENNIAL

Grateful acknowledgment is made to W. W. Norton & Company, Inc. for the permission to reprint eleven lines from "For an Album" from *Time's Power: Poems 1985–1988* by Adrienne Rich. Copyright © 1989 by Adrienne Rich. Reprinted by permission of the author and W. W. Norton & Company, Inc.

Certain names in this book have been changed. The altered names are Lenny Klein, Jess Marcus, John Feeny, and Special Agent Anderson.

A hardcover edition of this book was published in 1998 by Random House, Inc.

P.S.™ is a trademark of HarperCollins Publishers.

HarperCollins books may be purchased for educational, business, or sales promotional use. For information, please e-mail the Special Markets Department at SPsales@harpercollins.com.

First paperback edition published in 1998 by Harcourt Brace & Company.

FIRST HARPER PERENNIAL EDITION PUBLISHED 2010.

The Library of Congress has catalogued a previous edition as follows:
Shapiro, Dani.
    Slow motion : a true story / Dani Shapiro.—1st Harvest ed.
       p. cm.—(A Harvest book)
    Originally published: New York: Random House, © 1998.
    ISBN 0-15-600847-5
    1. Shapiro, Dani—Family. 2. Women novelists, American—20th century—Family relationships. 3. Parents—Death—Psychological aspects. I. Title.
    PS3569.H3387Z47   1999
    818'.5403—dc21   99-15150

ISBN 978-0-06-182669-6 (Harper Perennial edition)

19 LSC (C) 10 9 8 7 6 5

Our story is of moments
when even slow motion moved too fast
for the shutter of the camera:
words that blew our lives apart, like so,
eyes that cut and caught each other,
mime of the operating room
where gas and knives quote each other
moments before the telephone
starts ringing: our story is
how still we stood,
how fast.

— Adrienne Rich
"For an Album"

SLOW
MOTION

CHAPTER
ONE

The night before I receive the phone call that divides my life into before and after, my face swells in an allergic reaction to a skin cream, then blisters and chaps. I am at a health spa in Southern California, a place where wealthy older women go to rest and rejuvenate, where young matrons snap their bodies back into shape after pregnancies, where movie stars stretch out on massage tables in private Japanese gardens, offering their smooth backs to the sun.

I am none of the above, and for the past three days, since arriving at the Golden Door, I have often paused amid cacti and rock gardens to wonder what, exactly, I'm doing here. I am twenty-three years old, and my life has become unrecognizable to me. I have slid slowly into this state the way one might wade into an icy lake, dipping a toe in at first, then wincing, pushing past all resistance until the body is submerged, numb to the cold.

When the phone interrupts my post-hike breakfast of a half-grapefruit sweetened with honey, I am sitting cross-legged on my bed, listlessly flipping through the pages of the *San Diego Herald,* staring out the sliding glass doors at my private patio. I am upset about my face, which is itching and beginning to blister. My eyes are slits. I have never been allergic to anything before, and am worried that this rash might spread down my neck and across my chest, causing me to swell inside, my body choking on itself.

"Hello?"

"Dani, it's Aunt Roz, darling."

"Hi, Roz," I respond, confused. This aunt, who lives in sub-urban New Jersey, is not someone to whom I'm particularly close, and she would have no reason to know that I am at this health spa, much less track me down here at the crack of dawn. Though it doesn't occur to me to be frightened, though no alarm bells ring in my mind, I watch as my thighs begin to shake for no apparent reason.

"Dani, I'm calling because—"

She pauses, speaking very slowly, as if to an imbecile.

"The first thing you should know is that everything's all right," she says. And then, "Mother and Dad were in an accident."

"What kind of accident?"

"In their car, they—"

"Where were they? Where are they? Why are *you* calling me?"

"Now, Dani, if you'll just slow down—"

She keeps repeating my name, and she says it the way I hate, the way my mother's family has always said it, with a sort of pseudo-classy soft "a," as if we're from England, not New Jersey. There is an edge to her voice, as if she's somehow holding me accountable for being on the other side of the country at a moment like this. She thinks I'm a fuckup, a college dropout, a high-class drifter.

"They're both in intensive care," she says.

"Where?"

"Overlook Hospital, in Summit. They were driving home from your mother's office last night—"

"Last *night*?"

"It was late—there was nothing you could have done—"

I file this away somewhere, under miscellaneous family insanity. I am my mother's only child. My father has a daughter from his first marriage, my older half sister, Susie, who lives in New York City.

"Has someone called Susie?"

"No."

Jesus.

"How did you find me?"

"Your mother gave me the name of the place you're staying."

"So she's conscious—"

Aunt Roz snorts, actually snorts into the phone.

"Dani, your mother has two badly broken legs. Her tibia, her femur—"

Roz is a doctor's wife—the kind who thinks her marriage license includes a medical degree. Her husband, my uncle Hy, is a surgeon, and my favorite family member. I may be speaking to the wrong person.

"Where's Hy? I want to talk to Hy," I say. My voice has begun to shake along with my legs. Hy will tell me the truth. His hoarse, pipe-smoking voice will soothe me, tell me this isn't as bad as it sounds. I look wildly around my room at the sliding Japanese screens, the elegant, lacquered breakfast table upon which a fan has been set, detailing my day's activities: 9 A.M. aerobics, 10:30 stretch 'n' tone, 12:00 massage.

"Uncle Hy is with the doctors."

"How's my father?"

"He'll be fine—" Roz says flatly. "Not a scratch on him—and he wasn't even wearing a seat belt. It's your mother you should be worried about."

I don't stop to wonder why, if my father is fine, he isn't the one calling me in Southern California. My brain has gone numb, my instincts taking over. I will find out what has happened to my parents one small, manageable blow at a time.

"I'll get the next flight home," I say, calculating how long it will take to get to the San Diego airport.

"Good idea, Dan," says Roz.

I sit on the edge of the bed and dial a number in New York. There is a high-pitched buzz in my head: sounds, thoughts, language itself distilled into a single note of terror. I float out of my body and watch myself from a corner of the ceiling; this is something I do often—watch myself as if my life were a movie, as if I were only acting a role in this moment, as if it can be played back, cut, edited later.

His office phone rings once, twice, then is answered by his secretary, Marie, a woman who knows me—and my role in his life—well.

"Mr. Klein's office—"

Her voice is low and sexy, modulated within an inch of its life.

"Marie, it's Dani."

"Dani, how are you? How's California?"

I have often wondered how she keeps it all straight: wife, daughters, girlfriend.

"Is he there? It's an emergency."

She puts me on hold and I close my eyes, try to quiet the buzzing in my head. My heart is skipping beats, thumping irregularly in my chest. Years from now, when this happens, I'll wonder if I'm having a heart attack. But at this moment in my life, at twenty-three, I think I'm indestructible. I figure I have until I'm thirty. At thirty I'll expire, like a bright flame burning itself out.

"Hello, cupcake."

Lenny's voice pours through the phone lines, across the country, into my ear. The last time I spoke to Lenny was three days ago, when we were staying in Los Angeles at the Bel-Air Hotel, and we had such an ugly fight that I asked him to leave me alone for a few days—and he actually did.

"Lenny, something bad's happened."

"What?"

I can almost hear his head clicking off the possibilities. Something bad could mean virtually anything at this point: police, drug bust, vehicular manslaughter, God knows what.

"My parents—were in a car crash—it sounds pretty serious—"

The words are coming in gulps of breath. Saying it out loud, saying *parents* and *car crash* in the same sentence, saying it to Lenny Klein—it's all too much. My system is shorting out, and suddenly I'm panting.

"Honey, do you have any kind of bag," Lenny says gently. "A paper bag, a plastic bag—"

"What for—"

"Just do what I tell you. Do you have the bag? Hold it over your nose, and take some slow, deep breaths. In and out . . . in and out. Good girl."

I hold a piece of Saran wrap with bits of grapefruit still clinging to it over my nose and mouth, improvising a bag, and try to do as I'm told. I catch a glimpse of myself in the mirror above the bureau: I'm in gray sweatpants and a sweatshirt, my hair pulled back in a ponytail, my face red and swollen.

"I've got to get back," I say. "Could you have Marie get me on the next flight out of San Diego to Newark?"

"Easier said than done."

"What do you mean?"

"There's a fucking blizzard in New York," says Lenny. "They've closed the airports."

I grew up in an Orthodox Jewish home, a home where Sabbath was observed, my father wore a yarmulke, and we kept meat and dairy separate, according to religious dietary laws. Though I've strayed far from that home, in moments of pain, or shock, Hebrew words fly into my mind like a flock of blackbirds, foreign and unintelligible. They ride the crest of memory—these words and prayers—a whole other language I once spoke so fluently I even thought in it, and now no longer understand. Sometimes I think I have locked it deep inside myself and thrown away the key. Other times, I think it's accessible if only I know where to look: a language within my language, a heart within my heart.

So when I get off the phone with Lenny and dial my half sister's office number in New York, there is a tune drifting through my head, a prayer sung at the beginning of Sabbath services. *Avinu Malkenu, Harenu v'anenu* . . . I have not attended shul since leaving for college six years ago, at seventeen, but no matter. I can identify the song, sing every syllable the way, as a teenager in the 1970s, I knew every Springsteen lyric.

Susie, a psychoanalyst, is in session. Her machine picks up, and for a split second I almost blurt it out—*Dad and Irene were in a car crash*—but then I think of my half sister sitting in her office in Greenwich Village, surrounded by the accoutrements of her life: volumes of Freud, Oriental rugs, framed Ferenczi letter, burgundy velvet analytic couch. I picture her wearing her granny glasses and ethnic jewelry, her long wavy blond hair almost to her waist, a patient lying on that couch. I say it's urgent, to call back the minute her session ends.

If I am twenty-three, Susie is thirty-eight. She is a grown woman, certainly more grown than I am. She is an esteemed shrink, author of a book on schizophrenia, exotic traveler, and recently divorced from her psychiatrist husband. Her life, at least compared with mine, is sane and stable. Still, I somehow feel protective of her. I want to hold back the tide. She would say it is projection—that it is myself I am trying to protect here, flinging up my arms, shielding my face from the shards of a life swirling around me like broken glass.

I try to imagine my parents, but have virtually no information to go on. Which car were they in? My father's little sporty Subaru? My mother's Audi 5000? Where did it happen? What, exactly, happened? How is it possible that I don't know, at this very moment, whether my parents are alive or dead, in critical condition or just a bit banged up? Lenny said there's a blizzard. Was it the weather? Did their car skid off the highway? Were there other cars involved? I squeeze my eyes tightly against it all, but the images churn, they don't stop. I want a drink, a pill, anything. I methodically dig my nail into the palm of my hand. I want to move the hurt. I don't know how long I sit there—a minute? an hour?—before the phone rings again.

"Dan-Dan?"

It's my uncle Hy. Talking to Roz made me numb, talking to Lenny made me hyperventilate, and talking into Susie's answering machine made me mute. But hearing Hy's voice, filled with love, and with something else—something I can't yet identify—makes me weep.

"God, Hy, what's going on?"

"I don't know," he says quietly, this man I have always counted on to know. I decide to take it a parent at a time.

"How's my mother? Roz said she has broken—"

"Dani, your mother may never walk again."

A door slams shut inside me, then another, then another.

"And my father?"

"We don't know what's wrong with your father."

"Where is he?"

"He's in a coma, Dani."

"What happened?" I whisper.

"He passed out at the wheel. It may have been a stroke—we just don't know."

I finally recognize the unfamiliar note in Hy's voice: he's treating me like an adult, telling it to me straight.

"Get home," he says. "Get home now."

It's ten in the morning. My best bet seems to be the red-eye out of LA, which doesn't take off for another thirteen hours. Lenny arranges for a limo to take me from San Diego to LA, and has Marie book me on three afternoon flights to New York, just in case. I even try Boston and Washington, but it seems this blizzard is blanketing the entire East Coast. Lenny Klein is a man who can make virtually anything happen—that is, anything money can buy: center-court seats at the U.S. Open, oceanfront villas on the Côte d'Azur, Cuban cigars that arrive each month wrapped in plain brown paper—but he can't cut a path through a snowstorm. He has been known to charter private jets when commercial airlines didn't conform to his schedule. But all the money in the world won't get me home faster. No planes are landing anywhere, period.

The limo—stately, elegant, dark blue—pulls up to the gates

of the Golden Door. Everything Lenny does falls just to the south side of flashy. It's the mid-1980s after all, and flash is in the air, but as my mother has put it, Lenny has *taste*. "After all, he chose you," she has said more than once, placing me in the same category as his vintage Ferraris, his homes in Bedford, Jamaica, Martha's Vineyard. Lenny is a collector of fine things, and I am a thing. A girl in Lenny's girl collection. I guess my mother feels that if I have to be carrying on with a married man, at least I'm doing it with somebody rich and powerful, somebody who will show me the world.

I fold myself into the back of the limo, my single piece of luggage in the trunk. In the back of the limo there are assorted tapes, a sound system, and a telephone. What there is not—what I had been secretly hoping for—is a crystal decanter filled with something amber: scotch, brandy. I need to sedate myself for the three-hour drive. My body has not stopped shaking. I think about my mother and shudder at the degree of impact it must have taken to snap a thigh bone in two. Try as I might to imagine her with shattered bones—Hy mentioned both legs, pelvis, ribs, nose—I have always seen her as indestructible. A therapist once told me my mother reminded her of Mary Tyler Moore in the film *Ordinary People*. She is angular, energetic, fiercely private, imperious. How is she handling being flat on her back, in traction, at the mercy of doctors, nurses, and order-lies? Is she telling them what to do? My mind zings back and forth between my mother and my father—the words *coma, femur, critical, stroke* forming, dissolving, then forming again—as we pull away from the Golden Door and head north to Los Angeles.

I summon up my nerve. "Is there any scotch back here?" I ask the driver. The bright morning sun is muted by the tinted windows.

"No, ma'am," he answers, then pauses. "Would you like me to stop at a liquor store?"

"No, that's all right, thanks."

It is the first time in my life I have been called "ma'am."

The limo speeds north on the San Diego Freeway, and the fact of motion itself is a relief. We are going *somewhere*—moving fast in the wrong direction. I should be heading east on a plane right now, circling above Newark airport, waiting for a break in the sky, a small ripped-open seam that would allow us to land. I snap a Carly Simon tape into the cassette deck, lean my cheek against the cool dark leather as the car fills with the opening drumbeats of "You're So Vain."

The car phone rings. It's Susie. I had given her the number when she called me back at the spa.

"Dan, it's me."

Months sometimes go by without any communication with my half sister. I have no idea what she thinks of me—if she thinks of me at all. I always feel small around her; small, and stupid. I have looked up to her all my life. On top of being a shrink, she's a serious classical pianist who has studied for many years. I have listened to her play Liszt études and Chopin nocturnes, watched her graceful fingers flying over the keys, her brow furrowed in concentration as she transformed notes on a page into something that moved me. I have also studied piano since I was a child. I wanted to be like Susie. I have perfect pitch, and the music always came easily to me, but by the time I was in high school the last thing I wanted to do was spend long hours practicing alone, and so I tried out for the cheerleading team instead.

"Where are you?" she asks.

I look out the window at the arid landscape of Southern California. Heat waves rise up from the blacktop of the freeway. Billboards advertise condominium developments with names like Hacienda del Mar.

"I don't know," I answer. "Where are *you*?"

"I just got to Jersey."

"What's going on?"

"Well, Irene's going to be okay. She looks awful—like a caricature of someone who has been in a car crash—but Dad's in pretty bad shape."

Susie's voice is harsh and flat. As usual, my half sister minces no words. She's furious that no one thought to call her until I did—that my mother's family seems to have forgotten that she exists. The trouble between Susie and my mother goes back nearly thirty years to the time my parents first started dating each other. *She's a phony*, nine-year-old Susie told my father. *Don't marry her.* They have barely tolerated each other over the years—each has wished the other would disappear; my mother could be at death's door and Susie would probably perceive her as all right.

"I called Shirl and Harvey," says Susie, referring to my father's younger sister and brother. "I think they'd better get here."

I squeeze my eyes shut.

"Do you think—?"

I can't bear to say the words. Since I've been old enough to contemplate loss, I have imagined losing my father. Whenever we have been together and said good-bye, I have wondered if that good-bye would be the last. Though he is a bear of a man, an imposing figure, really, I have seen him as physically fragile, vulnerable, picturing a fatal heart attack, an embolism, a stroke—my father falling like an old heavy tree to the pavement on Wall Street, where he works, or while walking to temple on a Shabbos morning.

"The doctors are asking me what medication Dad's on," says Susie.

I think of the possibilities: Valium, Percodan, Codeine, Empirin. My father pops painkillers like Tic Tacs; he has suffered from chronic back pain for as long as I can remember. In the center of our breakfast table back home, on a lazy Susan where most people might keep cereals he has collected an impressive array of plastic bottles, each prescription written by a different doctor.

"I don't know what he's taking," I say. "Why don't you ask Irene?"

It is a function of my relationship with my half sister that I call my own mother Irene in her presence. I am trying to ally myself with her, to let her know that I understand.

"Irene's stoned," says Susie. "They have her on painkillers up the wazoo."

My head feels as if it's going to explode. My mother stoned is another in a series of impossible images. My father, comatose. My psychoanalyst half sister, my father's sister and brother, and my mother's suburban New Jersey relatives convening in a hospital corridor, pretending to get along. The rifts between my mother and my father's side of the family are deep. They go back at least ten years, to the time my father's sister, Shirl, had the flu and didn't attend my Bat Mitzvah. My mother was certain Shirl didn't have the flu and wasn't coming because she didn't think the service would be religious enough. The night before my Bat Mitzvah, my mother called Shirl psychotic and hung up on her.

"What's the weather like?" I ask Susie.

"Sucks."

"Am I going to be able to get home?"

"Eventually."

I close my eyes and count to ten. I want to scream at her, tell her to stop answering me in monosyllables, that I'm her sister, not her patient. I want to cry out for help—to let her know that I'm only pretending to be a grown-up, that in fact I'm a complete and total mess. But perhaps she knows this already.

"I'd better go," she says. "The doctor just came out of the ICU."

"Okay."

We are both quiet for a moment.

"Susie?"

"Yeah?"

"I love you."

We are two only children, raised by different mothers, fifteen years apart. Half sisters, connected to each other by half promises and half lies. But today we are all each other has in the world—and the man who connects us is fighting for his life.

I think she says "I love you too," but there is static, and we are disconnected.

My parents had three previous marriages between them. My father married Susie's mother when he was in his early twenties. It was a marriage that worked on paper. Early photographs of the two of them show a young, happy if slightly baffled-looking couple on the beach in Miami or playing shuffleboard at resorts like Kutchers and Grossinger's. But underneath her proper Orthodox surface, my father's first wife was a rebellious, intellectual spirit, and he had no idea what to do with her. Where he came from, women didn't aspire to more than a comfortable family life and perhaps some volunteer work at the temple sisterhood.

After Susie was born, the couple stayed very involved with both sets of in-laws, spending Shabbos dinners either at my father's parents' house on Central Park West or at his in-laws on Fifth Avenue. My father was on the road half the time, traveling to a small town in Virginia, where he was overseeing the family silk mill. Years later, my grandfather would shut down the mill for good, and lend my father the money to buy a seat on the New York Stock Exchange. But back then, my father's traveling must have taken its toll. When he was home, he and his wife fought viciously, or lapsed into tense silences. Still, my father may not have grasped or understood her growing frustration and disenchantment.

When Susie was six, my father returned from a business trip to find an empty apartment, with only his clothes left hanging in the closet. His wife, daughter, and all their belongings were gone. There were rumors, of course, that she had run off with Susie's pediatrician. This kind of thing just didn't happen. In Susie's class at Ramaz, an Upper East Side yeshiva for girls, she was the only child of divorced parents.

My father waged a custody battle for Susie, and won ample visitation rights: Wednesday nights, every other weekend, and Jewish holidays. In the meantime, he was probably being fixed up on blind dates all over town. He was already becoming a tragic figure of sorts, ditched abruptly by his flighty, good-for-nothing wife. On the weekends he had Susie he sometimes took her to resorts in the Catskills, where they'd play a game: he'd go out on dates with young women, and Susie would narrow her little six-year-old's eyes and give him her opinion.

For his second wife, he chose another daughter of a privileged Orthodox clan. Dorothy Gribetz was a lovely, sweet-natured girl, and according to everything I've ever heard, my father was crazy about her. So was Susie. He proposed, she accepted, and plans for a wedding were set in motion. It wasn't until a short time before their wedding that my father's best friend told him a rumor that had been whispered throughout the Orthodox community: Dorothy had Hodgkin's disease, which in those days was a terminal illness. She didn't know it—her parents had kept it secret from her, and from my father as well.

A few nights before their wedding my father paid a visit to Dorothy's father. Was it true? Was she dying? Yes, he told him, Dorothy's prognosis was that she had a year to live. He had kept it from my father because he saw how happy he made Dorothy, and he wanted her to have that happiness, even if only for a short time—even at the expense of my father's, and of Susie's.

I picture my father now, standing beneath the *chuppah* on his wedding day. He is not the father of my memory, but of my imagination: he is a young man—perhaps he is thirty-two—but his eyes are already old. He turns to watch his bride walk down the aisle. She is a vision of innocence in her simple white gown. This should be the happiest day of his life. His eyes sting as she moves toward him, flanked on either side by her parents, and his heart is hollow. He is watching himself become a widower. He looks around the shul at the assembled guests and blinks hard against the thought that they will all be gathered here again in the not too distant future, that he and his bride will be together one last time in this sanctuary: he in the torn black clothes of mourning, she in a plain pine box.

I grew up absorbing my father's sadness without knowing where it came from. Sometimes he just disappeared. Not like other fathers—fathers I heard about, who drove off in their cars and never came home again—but just faded, as if he couldn't really be there, not all of him. He would be sitting in a lawn chair smoking a Camel, and all of a sudden his eyes would grow vacant, his mouth would crumble, and he would stare off into the distance. I would follow his gaze to see what he was looking at, but I never saw what was making him so sad. I couldn't make out the faint shadow of his first wife against the forsythia hedge in the backyard, holding a little-girl version of Susie's hand. I couldn't see Dorothy huddled in a blanket by the seashore, weak and pale in the final months of her life.

There was defeat in the stoop of my father's shoulders, or in the way he shook a few pills into the palm of his hand, then downed them in one gulp when he thought no one was looking. I thought that perhaps this was what it meant to be a grown-up; that along with growing big and tall, the pinprick of sadness that was inside me too would spread until it covered my insides like a stain.

I was sixteen years old before I heard about Dorothy. Susie let it slip—*when Dad, Dorothy, and I were upstate one time,* she said—and when I looked puzzled, she stopped and stared at me. *You don't know about Dorothy?* Susie was by then a thirty-one-year-old psychoanalyst, and on some level, she must have known what she was doing. Perhaps she felt I needed to know. There were already danger signs—signs that I was fading fast myself.

And then there was my mother. Over the years, the story of my parents' courtship and marriage has acquired a delicacy that has kept me at a distance, like an ancient hand-blown piece of glass that might disintegrate if I got too close.

What I was told as a child was this: they first met on East Ninth Street in Manhattan, where they were across-the-street neighbors. It was a Saturday, the Shabbos, and my father was walking home from shul with nine-year-old Susie. My mother was returning from the hardware store, where she had just bought a hammer. Hammers and Shabbos are two things that don't go together: for Orthodox Jews, the Sabbath is a day of rest, when no work is to be done, certainly not manual labor. So when my father met my mother, he must have known she wasn't from his world.

It is a story my mother has often recounted feverishly, slam-dunking the metaphor of the hammer. *He knew I wasn't obser-vant. He saw the hammer. He knew what he was getting himself into.* She was a career girl with her own advertising agency, who had left her first husband when she was thirty. My father was so taken with my mother that, the following Sunday, he pored over the Manhattan phone directory, searching for her. He knew only her first name—Irene—and her address on East Ninth. I can only imagine what he was thinking, the way his heart must have been racing. Who was the dark-haired beauty from across the street? She looked to be in her early thirties. What had she been through? Why wasn't she wearing a wedding ring? Why was he tracking her down despite his better judgment? He ran his finger down each column in the White Pages, looking for *Irene*s or the initial *I* on East Ninth Street until he finally found her in the F's, under *Fogel,* a surname left over from her first marriage.

In a photograph of my parents I have hanging over my desk, they are walking down the aisle of Young Israel of Sixteenth Street on their wedding day. My father is dashing in a well-cut dark suit, and my mother is elegant in ankle-length ice-blue. Their arms are linked as they walk together toward the *chuppah,* and my mother is smiling triumphantly at whoever is taking the picture, a thin cloud of netting floating over her face. My father is smiling too, but now, if I look beyond the smile, I see that he is haunted. There are ghosts in his mind, ghosts swirling all around my father and my mother in the moment before they take their vows.

I am not yet born, and there is already a piece of my father that is dead.

I am drunk, halfway home.

Or rather, I should be drunk, but nothing seems to be working: not the two vodkas I had in the airport bar, not the bad airplane wine I have been drinking since takeoff. I've recently reached a point in my drinking where one drink can get me drunk or ten can have no effect. But it isn't tipsiness I'm after. I'm looking to anesthetize myself from head to toe, which is why I'm mixing red wine and vodka—a very bad idea and I know it. All I want to do is stop feeling. I want the images of my parents in my mind to fade until there's nothing but a warm, sickening haze, until I get just dizzy enough to pass out in my seat.

I always drink on airplanes—I consider them a sort of time-free zone, an endless cocktail hour. Besides, I'm terrified of flying, and after a few little bottles of Smirnoff and cans of Mrs. T.'s Bloody Mary mix, I can usually forget that I'm in the air, at least for a little while. Forgetting is what it's all about—forgetting that I'm twenty-three years old and have nothing to show for it. Once I've had a few drinks I can convince myself that I have a lot to show for it. Who needs things like college degrees, nice hometown boyfriends, starter jobs at advertising agencies? My friends are all playing a game, and I have stepped to the sidelines. I have chosen to sit this one out.

Instead, I am playing house with Lenny, zigzagging across the country at his beck and call. I have something resembling a career, halfheartedly modeling and doing television commercials. For the moment, I think I want to be an actress. I dropped out of college three years ago after being cast in a York Peppermint Patties commercial, and now I feel that I'm stuck with it—

acting and Lenny—as if, having taken a wrong turn, I have had to make a commitment to follow this road wherever it takes me. Retracing my steps has not felt like an option. I have run faster and faster in the wrong direction, eyes squeezed shut, hoping that somewhere along the way the road will loop around again.

A red-faced, middle-aged man is sitting next to me, matching me drink for drink. I've noticed him sneaking glances at me. A thick annual report is spread over his tray table.

"I'll bet you're an actress," he says. "Am I right?"

"Right," I say faintly.

"Have I seen you in anything?"

I reel off the list of my most recent commercials. Hess Gasoline, Coca-Cola, Scrabble. My words are slightly slurred. Although I think I don't show it, I am always flattered and surprised when people ask if I'm an actress or a model. As far as my looks go, I am seething with insecurity, a bottomless pit into which compliments fall for a brief, shining moment, then disappear. The whole notion of physical beauty has grown increasingly important to me as my intellectual curiosity has vanished. A few years ago I was studying music and literature at Sarah Lawrence, diagramming Mozart concertos and reading Tillie Olsen. But why struggle with a term paper on the elements of foreshadowing in *Bleak House* when I could be cavorting on the beach in front of a camera and getting paid for it? Why deal with caked-over tubes of toothpaste, smelly refrigerators filled with old cartons of labeled food and turned milk when Lenny Klein has handed me keys to an apartment high above Central Park South? I have used myself as a physical instrument, slicing my way through the world with nothing but youth, long legs, and long blond hair. At times I think I have chosen the easy way, but every once in a while I realize that this may be the hardest way of all.

"So, are you heading to the big city on business?" my traveling companion asks.

"No."

I pause and take a gulp of wine.

"My parents were in a car crash. They're both in intensive care."

"Sorry to hear that," he says, recoiling slightly.

"I actually don't know if they're alive or dead. It looks like my father had a stroke while he was driving."

I look at the bloated lines of his jaw, his thick hands, a gold insignia ring encircling his stubby pinkie finger.

"Tough luck," he says. "But you'll get through."

"I can't believe this is happening," I murmur, more to myself than to him.

"Yeah. Well, shit happens," he says.

We bump through the air, and the FASTEN SEAT BELTS sign lights up with a *ding*. Usually this would be enough to send me into a panic, but not tonight. What are the odds of a car crash and a plane crash in one family in one day? I turn away from my traveling companion and cover myself with a thin airline blanket. My wineglass is empty, I wedge it between an airsickness bag and an in-flight magazine in the pocket in front of me. *Shit happens.* Is there some sort of hard-won middle-aged wisdom in that notion?

The captain announces that we're experiencing some turbulence. I curl up in my seat, trying to find a quiet place in my mind where I can rest, if not sleep. I'm afraid to drink any more. My head is spinning, and each lurch of the plane turns my stomach. Lenny has arranged for a car and driver to pick me up at Newark, and I'm planning to go directly to my uncle Morton's house in Summit, only a few blocks from the hospital.

I try to conjure up Lenny's face, but he fades in and out of focus: a thatch of dark hair, big brown eyes, a thick wrestler's body gone soft around the middle. His most expressive feature is his voice, which is deep and raspy, a tool he uses to great advantage in the courtroom. Lenny is a name partner in one of the largest law firms in the city—a firm whose other partners include three former U.S. senators. *Fortune* recently listed him as one of the top five litigators in the country. I suppose he also uses his physical self as a tool—striding, pausing dramatically, rolling his eyes, raising his voice to a thunderous pitch or lowering it to a whisper. He used that voice to seduce me three years ago.

I met Lenny Klein at Sarah Lawrence. He was the stepfather of one of my close college friends. The first time he called me, that hoarse voice asking for me on the dormitory phone, he said he wanted to get together, something to do with Jess. Would it be absurd to say I believed him? There must have been an odd feeling in the pit of my stomach, but I ignored it. I agreed to meet him one evening in the city—and I agreed not to tell Jess. Her birthday was coming up; I thought maybe he wanted my help in planning a surprise party.

And when he picked me up on a prearranged corner in his white Rolls-Royce, and his arm slid familiarly across the passenger seat, just brushing the back of my neck, what was preventing me from opening the car door and getting out? I was twenty years old, and the idea that a friend's father—a friend's *married* father—would try to seduce me was something I found unfathomable. And yet, at the same time, I felt thrust into a parallel universe, one I had never known existed. Everything I knew about right and wrong seemed to vanish inside that car.

The truth is, Lenny repelled me before he attracted me. I went through the motions that night, let him take me to an elaborate dinner where I consumed the better part of two bottles of wine, but I had no intention of ever seeing or speaking with him again. I would have some explaining to do to Jess—or maybe she didn't need to know. After all, I thought that would be the end of it. A little adventure, an honest mistake. And even when he started calling me several dozen times a day, even when he drove to my dorm and parked his car outside, I held my breath, just praying Jess wouldn't walk by. He sent me flowers and cards— the floor of my old dorm room was covered with vases of yellow roses, the constant faint scent of decay in the air.

Another twenty-year-old might have called the campus police, filed a complaint. But secretly, Lenny's attentions made me feel like the most special girl in the world.

The in-flight movie is *Gorky Park*—a film I auditioned for a couple of years back. I watch the tail end without headphones. The actress they cast looks a whole lot better as a Russian spy than I ever would have. I try to focus on the screen, but I'm seeing double, so I close my eyes. Beneath my lids, another film is taking place: My parents' Audi collapses like an accordion against a concrete highway divider, my father's head is flung in slow motion into the steering wheel. His eyes close, glasses crack, lenses pop out from the impact. My mother screams, an unearthly sound, as her legs are mangled beneath her. Steam pours from the hood. All around them, giant flakes of snow drift silently across the nearly empty highway.

I open my eyes, blink hard, and gasp for air.

"You all right?" my neighbor asks. He has moved from wine to Baileys Irish Cream, and his face is the color of sunset.

"Yeah," I lie. "I'm fine."

I make my way down the aisle to the lavatory and splash cold water on my face, then examine myself in the mirror. The rash is getting worse. My cheeks are streaked with tears, and my lips and eyes are all puffy.

The captain's voice pipes into the restroom, announcing that we're about to begin our final descent into the Newark area. The weather in Newark is a bracing eighteen degrees and we should be touching down at approximately six-fifty, local time. Before I return to my seat, I meet my own gaze evenly. The words of the *Sh'ma*, a Hebrew prayer, tumble through my mind.

You are alone in the world, I whisper to the poor, pathetic girl in the mirror, preparing for the worst.

CHAPTER
TWO

Newark airport. The crack of dawn.

My breath tastes like sour milk, and there is an
acidic tightness in the back of my throat. My
sinuses ache. I half run, half walk to baggage claim,
where Lenny told me the driver would be waiting.
My mink coat is flung over my shoulders, and the
heels of my black boots, which are worn down to
metal nubs, make sharp, staccato clicks against the
airport linoleum.

As I run through the near-deserted terminal,
past closed duty-free shops and newsstands, my
heart slams against my chest. Now that I'm here,
only a half hour's drive from the hospital, it's as if
I'm being pushed from behind. Nothing is happening fast enough. I want to run through the
sliding doors, head north on Route 22, jog all
the way to Route 24 which leads to Summit. I will
be a speeding black blur, a cartoonish smudge
against banks of freshly fallen snow. I will get to
my parents under my own steam, and then I will
do whatever is necessary. I will knit my mother's

bones back together. I will kiss my father's cheek and he will wake up.

Even though Lenny did not say he'd meet me at the airport, I scan the bank of drivers holding up handwritten signs in the baggage claim area, not looking for my own name, but searching for Lenny. After all, shouldn't he be here? Isn't this a big enough deal? It occurs to me that I didn't ask Lenny to come. I wasn't sure I wanted him to accompany me to the hospital, where I would have to explain his presence to my relatives. *Aunt Roz, Uncle Hy, this is Lenny Klein* is an introduction, all things considered, I'd just as soon never make. Not to mention Uncle Harvey and Aunt Shirley, my father's Orthodox brother and sister, who barely accept the idea that I don't keep kosher. But a married man, twenty-three years my senior? Lenny makes eating a bacon cheeseburger look like a mitzvah.

Finally I see a sign with my name on it. Well, not exactly my name. Lenny has an embarrassing nickname he likes to call me, and he's decided to be funny: a uniformed driver is holding up a sign that reads *Ms. Fox.*

I tell the driver about my parents' accident. He heard about it on the radio. It must have been pretty bad, he says. The police closed off a whole section of Route 24 near the Short Hills Mall—they do that only in the worst cases. He asks me what happened, eyes darting at me in the rearview mirror, but I shake my head. I have been awake and traveling for twenty-two hours, and I still don't know.

As we leave the airport, traversing the loose gray tangle of highway ramps, I breathe in the acrid, familiar scent of northern New Jersey. These oil refineries, suburban malls, and the jagged, graffiti-covered cliffs of Jersey City are the landscape of my childhood. The neon Budweiser eagle slowly flapping its wings over the Anheuser-Busch plant in Newark has always signaled a return home.

Snow is piled along the sides of the highway in great big banks, and the few cars negotiating the roads at this hour are moving very slowly. There is an abandoned Volkswagen near an exit ramp, its round roof covered with at least a foot of powder. The blizzard did not end until the early morning hours, and even in the weak light of dawn, it is unbearably, monotonously white. I feel as if I have returned to the country of my childhood only to find it altered, subterranean.

We pass the exit for Hillside, where I lived for the first seventeen years of my life. Hillside is a small town near Newark and Elizabeth—the part of New Jersey the jokes come from. I grew up in a red-brick Georgian Colonial with white pillars and slatted shutters. On breezy days the putrid smoke from the refineries and landfills would waft over our flagstone patio, forsythia hedges, and kidney-shaped swimming pool. There were rumors of toxic waste and mysterious diseases. My parents kept the windows shut; the air could kill you. Even the air *inside* the house was regularly tested by men who went down to the basement wearing gas masks.

There were families who had lived in Hillside for generations, like the Peabodys and McCarthys, who drove old station wagons and wore threadbare cashmere cardigans. Then there were Jews who wished they were like the Peabodys and McCarthys, and who sent their daughters for nose jobs and their sons to golf camp. There was a smattering of concentration-camp survivors with thick accents and numbers burned into their wrists, families who were, for the most part, related to one another. The Wilfs and Resnicks had settled in New Jersey after the war and started building shopping malls. They had tennis courts in their backyards that were used by everyone in the neighborhood except them.

And then there was our family. We never quite fit in. My father's religious observance set us apart, although my mother tried to make a place for herself in the community, doing volunteer work, joining the temple sisterhood. She had a regular doubles game with three other women, and once a week, off she'd go in her tennis whites, the decorative pom-poms on the backs of her socks bobbing as she ran to the Cadillac or Lincoln tooting its horn in front of the house.

On Shabbos, when my father returned home from shul, a few of the neighbors with thick Eastern European accents would sometimes pay an afternoon visit. Sipping iced tea out of green plastic glasses, the adults would sit in the backyard on yellow-and-white striped chaises longues, while I'd stand on the toilet and spy on them from my second-floor bathroom window. Eventually my mother would call for me to come say hello, and I'd slink uncomfortably into the backyard. I knew what came next: *We could have used you in the camps, little blondie,* they'd say, patting me on the head. *The soldiers would have given you extra bread.*

In Hillside, I grew up alone in a room. I went to a yeshiva in another town; my school friends were a driving distance away. I was the only child in my mother's entire family—there were no cousins around, no children at all. I spent my early life surrounded by silence, thinking my thoughts, dreaming my dreams, inventing a self out of thin air. I had no one to reflect this self back to me. Not my father, who was already retreating behind a wall of pills and prayer. And not my mother, who dressed me in pretty clothes and treated me like a breakable doll. Our house was as still and quiet as a wax museum. Behind the closed door of my room, I wrote stories that I shredded when I finished and let float like confetti into the wastepaper basket, stories of a girl with brothers and sisters, a healthy father, a happy mother. I felt tangled up inside myself like a weed. But I didn't know what to do about it. It seemed my parents were tangled up too.

*   *   *

The driver points to some skid marks on the highway, and blue-black paint marks for several feet along the center divider. There are shards of glass still winking in the snow and orange-and-white barriers pushed to the side of the road.

"This must be where it happened," he says.

As we pass the scene of the accident, I crane my neck, searching for clues, for meaning. My eyes sting, and I open and close my fists, wishing for someone's hand to hold. I have never longed for a brother or sister the way I do at this moment. So as I move toward my parents, afraid of what I might find, I should not be surprised that I'm alone. Though I have a handful of good friends and a quasi-boyfriend, I am going through the most frightening moment of my life without solace, without witness. I make a bargain with God: *Let them live and I'll leave Lenny. Let them live and I'll go back to college. Let them live and I'll stop drinking. . . .* Hillside is a blur, just as it is in my memory. I try to picture myself as that girl standing by the open bathroom window, watching her parents on the chaises longues below. I hear the rattling of ice in green plastic glasses, and I see the woven disk of my father's yarmulke on top of his bald head. I have run so far and so fast away from that girl that I barely know who she is anymore.

Lenny's stepdaughter, Jess, and I met each other on Accepted Students Day at Sarah Lawrence. Neither of us had decided to go there—I was debating between Sarah Lawrence and Barnard, and Jess was waiting to hear from Vassar—but somehow, during that initial meeting, a silent pact was forged between us: *I'll go if you go.* We liked each other that much, based on nothing more than a first impression.

Jess had swingy brown hair cut into a pageboy and a thin, intelligent face. I thought she was beautiful. She was my physical opposite: Dark, angular, mysterious. I was sitting with my mother—it was Shabbos, and my father had not accompanied us to this event—and Jess was across in Reisinger Auditorium. Of all the students in that room, and there must have been two hundred of them, I focused on Jess Marcus. I barely noticed the man next to her.

After the business of selling the college to us was over, Jess and I met on a sprawling lawn where there was a reception for prospective students and parents. As I recall, she approached me.

She came right up to me and stuck out her hand.

"Hi, I'm Jess," she said. "I noticed you in the auditorium, and thought you were someone I'd like to get to know."

I couldn't imagine why. I saw myself as a formless, shapeless blob from New Jersey. I had gone to a yeshiva until seventh grade, and then to prep school; by the time I was looking at college, I was in the middle of a full-blown identity crisis. I was wearing an Indian-print skirt, an orange cotton blouse, and a matching Indian shawl—my stab at looking like a bohemian Sarah Lawrence type. Jess was cool, classy, her voice lilting and musical. I developed an instant crush on her.

Now, when I think of Jess, it is in still images, moments that remain frozen in memory. It is freshman year, just after winter break, and we are walking arm in arm to the pub after studying in the library. Jess is wearing a short down jacket, jeans, and boots. Somehow everything she wears looks elegant—even a down jacket. We order beers and sit in the corner at a small round table, our backs to the room. When I'm with Jess, I don't want to be with anyone else. Jess laughs at the boys who come over to talk to us, snubbing them with a sweep of her dark eyes. She touches me when she talks, her fingers resting on my arm for an instant or brushing my hair off my shoulders.

I have never had a friendship like this one, which borders on the romantic. I emulate Jess in every way. Her clothes, her voice, her sense of mystery. Of course I am nothing like her, but that's precisely why, at this moment, I wish I could blink and become her.

She is saying something to me, leaning forward, whispering. If I strain back through the years and quietly listen, I can almost capture her voice.

"That guy over there keeps staring at us," she says.

I turn around and glance at a kid on the burger line who does indeed seem to be focused on our table.

"He's looking at *you*," I say.

"No, *you*," she replies.

And with our heads bent together, our backs to the room, she looks at me queerly, as she often does, with a small Buddha-like smile. When I ask her what she's thinking, she just shakes her head at me with benign amusement, as if I were a beloved but slightly daffy pet. It is her secretiveness that fascinates me, that pulls me in. I feel like an open book, a fresh-faced, innocent little kid next to Jess. I want some of her sophistication to rub off on me.

Years later, Lenny would remind me that he introduced himself to me while I was talking to Jess that first day at Sarah Lawrence. *Hi, I'm Lenny Klein, Jess's stepfather,* he might have said. I have no such memory. If I remember Lenny at all from that first day, it is as a caricature: a stocky, middle-aged guy, a flash of gold watch, a pair of faded jeans that looked silly on his body, as if he really belonged in a suit. If you had told me back then that one day, in the not too distant future, I would be walking down a Paris boulevard with Jess's stepfather, that I would be lying in bed with him in a Cannes hotel room, that he would lean back in the upholstered chairs of haute couture salons, puffing on a cigar and choosing dresses for me—I would have laughed. I would have told you he was old enough to be my father, that he wasn't even cute, that there was no way.

Uncle Morton does not raise an eyebrow when I am deposited on his doorstep. He opens the door, looking tired and drawn in a silk bathrobe and pajamas, his breath clouding the bitter cold. He's been expecting me since yesterday. He takes in the hour, the limo, my one-piece black unitard more appropriate for dinner on Melrose than breakfast in New Jersey. He hugs me hard, strokes my head, says, *They're both alive*—then leads me into the kitchen and pours me a cup of coffee without another word.

Morton is my mother's older brother, and though I don't know him well, I feel at home with him. He looks like a male version of my mother, with the same almond-shaped eyes, high cheekbones, and regal bearing. How my mother and her siblings picked up this haut monde manner on a New Jersey chicken farm is beyond me. Morton has been married for many years to his third wife, Shirley Sugerman, a woman who has always been referred to in my family by her first and last name. *Morton is coming to dinner with Shirley Sugerman,* my mother might say, as if to differentiate between that Shirley and all others. Morton is a retired college professor and Shirley is a psychoanalyst, a profession my mother disdains, particularly in family members— which, in my family, is a problem. My half sister, Susie, also doesn't think much of Shirley Sugerman. Susie, who received a doctorate in clinical psychology, feels superior to Shirley, who received a doctorate in philosophy and then did psychoanalytic training. And my mother, who has a master's degree in social work, doesn't believe in psychoanalytic mumbo jumbo, but rather in the practical, behavioral tenets of family therapy. All this academic squabbling is over my head, of course. In a family of Ph.D.s, I am a college dropout.

"Visiting hours start at eight. You really ought to eat something," Morton says, popping an English muffin into the toaster oven. "Did you have anything on the flight?"

I wonder if he can smell wine on my breath. I bought Velamints at the airport newsstand and chewed half a package on the way here, not so much to mask my breath as to keep my mouth busy. I'm wired, as if I've been up all night snorting coke. My mouth feels tense, and my jaw aches.

Early morning light filters through the kitchen window, the sill piled high with snow. Icicles hang like daggers, and the branches of the old blue spruce in the backyard are sagging. One of the two Volvos in Morton's driveway is nearly plowed in, but the other has seen some action in this blizzard. I imagine Morton's receiving the phone call that his sister and brother-in-law were in the ER at Overlook, and making his way gingerly, disbelievingly, through the storm.

Overlook Hospital is on a hill, though what it looks over is nothing to speak of. There is a driveway and ambulance ramp in front, a large parking lot, scattered houses in the distance. The people who live in these houses must be used to the sound of sirens puncturing their nights. In LA, I once saw a woman make the sign of the cross behind the wheel of her car as an ambulance sped by on the freeway. Do these neighbors of the hospital say a prayer each time they hear a siren, or have they gotten used to it? When my parents were raced to the emergency room, were they taken in separate ambulances, one wailing behind another? A string of questions tightens in my head, stretching my brain to the snapping point, as Morton pulls up to the main entrance of the hospital.

"Go ahead in," he says gently. "I'll park the car."

I look at him out of the corner of my eye and realize he's thought this through, he's trying to do the right thing. No one can help me with this moment.

Morton pats my back as I open the passenger door, says he'll be waiting for me in the lounge.

This fear is unlike any I've ever known. It begins in my bowels and spreads through my heart, into my brain, a psychic brushfire—visceral, primitive. I'm terrified that I will see my parents and break into a thousand pieces. All around us, just to either side of this hill, the families of Summit, New Jersey, are stirring awake. Children, cozy in their Doctor Dentons, are dreaming of the snowmen they will build, with carrot noses and Oreo-cookie eyes. Life goes on as usual, and I am entering a parallel universe. Nothing is usual, and nothing can be taken for granted. With each step, I am moving closer to the end of my life as I have known it.

I follow signs pointing to the ICU. I hear popping sounds in my head like snapping wires. For years now, I have worried about my sanity. I have known that the combination of drinking, drugs, isolation, starvation, and Lenny could not possibly be good for me. I have even called Susie from time to time to ask if—in her professional opinion—it's possible for an otherwise healthy person to give herself a nervous breakdown. Susie always responds to this with a kind of analytic cool. She cites textbooks, the *Physician's Desk Reference*, the DSM-III. She seems not to think a psychotic break is in my future. Garden-variety neurosis, perhaps. A particularly rough late adolescence. *Would it make it easier if you were really crazy?* she once asked, offering a free interpretation.

My vision is shaky, as if I were seeing through a handheld movie camera. The gift shop, potted plants, orange plastic chairs in the lobby all look like pieces of a puzzle, but I can't quite see the whole picture. Passersby are looking at me, heads turning. I must look like a crazy woman with a face rash and dirty hair, reeling through the lobby in a panic, a spinning top, circling and wobbling until finally I end up at the glass-paned doors of the ICU, and I am shaking so hard I'm afraid to move.

I peer through the doors. The unit is a semicircle of curtained partitions with a nurses' station at its center. I see a few nurses in white, a bank of complicated-looking monitors, but beyond that I can't make anything out. As a child, I used to crouch by my parents' closed bedroom door, listening for sounds. I didn't know what, exactly, I was after—but I couldn't stop myself, each night, from tiptoeing down the carpeted hallway in my bare feet and flannel nightgown and pressing my ear to the keyhole. It seemed to me that my parents had each other and I had no one. They had secrets and stories—even their terrible fighting was a form of intimacy. Now, once again, I am standing on the other side of a door separating me from my parents. Are their beds next to each other in the ICU? My parents have always called each other *Wahoo,* a mutual nickname long held over from an Indian movie they saw early in their courtship. Can my mother whisper *Wahoo*? Can my father hear her?

I push the door open and walk directly to the nurses' station, focusing straight ahead. I want somebody to tell me how my parents are before I see them.

"I'm—I think my parents are here—" I say haltingly to the first nurse who makes eye contact.

"Shapiro?"

I nod. There are six beds in the ICU. I wonder how often they get a husband and wife in here at the same time. The nurse points to the second bed from the left, which is completely surrounded by curtains.

"Your mother is in there. She's been asking for you."

I inhale sharply. The air smells preternaturally clean.

"And my father?"

"Why don't you see your mother first," the nurse says.

Tears are rolling down my face in spite of myself. I swore that I'd hold it together, that I wouldn't fall apart until some time down the road. I've been pretending to be a grown-up for so long that for a moment I thought I actually was one.

The nurse reaches across the counter and pats my hand. I realize I've been gripping the edge so tightly my knuckles are truly white.

"Come with me," she says, "I'll take you over."

My mother is swathed in white bandages and casts from head to toe, and her legs are in traction, dangling above her like the tangled appendages of a marionette. Susie's words come to mind—*like a caricature of someone who's been in a bad car accident*—but I've never seen a caricature like this. Her head is turned to the side, away from me, her cheek against the pillow. I don't think she hears me approach her. I walk around the foot of the bed, my breath shallow, all my focus on keeping my expression impassive.

"Mom?"

Her face is bloated, one eye swollen shut. Her nose is black-and-blue mush, and a deep gash above her left eyebrow is stitched together, painted with an orange tincture.

Her right eye struggles halfway open. She fixes a watery gaze on me, rheumy and unblinking like a sick old dog's.

"I told them not to call you—" she whispers thickly.

Is she hallucinating? I note the IV slowly dripping into a vein on top of her hand. What is she on? My mother, who loathes Tylenol, is probably being fed six different kinds of painkillers and tranquilizers. She is too drugged to even know she is drugged.

"You had to come all the way from California. . . ." she murmurs, eyelid fluttering.

"Ssshhh—don't talk—"

I look for a place on her body to touch her. One hand is bandaged into the IV, the other is beneath the sheet. Both legs are in casts. Her color is an appalling yellow, like the ring around a bruise. I'm afraid to stroke her face, the fragile, shattered bones of her nose and cheekbone. When I was a little girl I used to crawl into the enormous bed she and my father shared, which was really two twin beds, and I would lie in the crack between the beds, picturing the place where the two beds met as a fault line, a fissure that could open up and swallow me whole, pushing my parents away from each other and me into the vast, dark, swirling cosmos. I would scoot to my mother's side of the bed, pressing myself up against her warmth until finally I would feel safe enough to fall asleep.

Now I want to crawl between all these wires and tubes and wrap myself around my mother. I want to hold her bones in place until they knit back together. I can't let her see the fear in my eyes. She is staring at me, her one unblinking eye filled with tears.

"Paul . . ." she mutters, her brow creasing.

"Daddy's just over there," I point through the closed curtains. "I haven't seen him yet. Do you want me to go see how he is?"

She nods.

"Okay. I'll be back in a few minutes," I say, rising from my knees, feeling the room buckle and swirl. I lean over the rail of my mother's bed and kiss her gently on the forehead, next to the stitched-up gash. Her skin is burning.

I make it to the other side of the curtain, then feel my hands rise up to my head. If I'm going to pass out, I'm in the right place. I try to take some deep breaths, but my throat closes up and suddenly I'm fighting for enough oxygen. *Panic*, I tell myself. *You're panicking*. I close my eyes and try to formulate a single, coherent thought.

"Are you all right?" the nurse asks, raising her head from a medical chart.

"My father—" My voice cracks.

She gestures to an open curtain in the semicircle.

"He's over there, but he's not awake."

My legs are rubber, but somehow they propel me to the foot of my father's bed. If my mother is yellow, my father is a pale, sickly green. He is lying flat on his back, his head rolled to the side, an oxygen mask over his nose and mouth. I hear a steady beeping, and realize that he's hooked up to an electrocardio-graph. My eyes dart to the peaks and valleys on the monitor, neon green on black. The rhythms of my father's heart look jagged. I feel my own heart speed up.

I bend over him gingerly, careful not to disturb the oxygen mask, and press my lips to the top of his head. My father has been bald since before I was born. His head has always seemed fragile to me, round and soft, slightly dented like a grapefruit. I inhale deeply, breathing him in. There are no casts, no broken bones. Except for his color, he looks like a man taking a midday nap.

I reach for his hand.

"Daddy?"

Nothing.

"Can you hear me?"

Not a flicker, a reflex, a blink. He may look as if he's sleeping, but the word *coma,* that terrifying word, focuses itself into sharp relief. How is it possible that my mother is shattered but conscious, and that my father barely has a scratch on him but is in a coma? Roz said my mother was wearing a seat belt, and my father wasn't. None of it makes sense. What happened when he passed out behind the wheel? Did he fall forward, his foot deadweight against the gas pedal? Did she try to unbuckle herself so she could reach across him and take control of the car?

"Ms. Shapiro?"

A man a few years older than me in a white lab coat materializes at my side. He introduces himself to me, and while I try to adjust to the idea that someone this young can already be a doctor, he lifts my father's eyelids and shines a pinpoint light into them. My father's eyes are murky, the color of a fish's underbelly, and his pupils are dilated.

"I want to talk to you about—" he begins, but I motion him out of my father's earshot. I've heard too many stories about people in comas who later repeat back whole conversations that took place while they were unconscious.

We walk into the corridor.

"I was in the ER when your parents were brought in last night," he says. "They're lucky to be alive."

He looks at me, this doctor, taking in my rash, my twenty-four-hours-and-counting rings under my eyes, my emaciated body in its black cat suit, and steers me across the hall into a staff lounge. He pours some orange juice from a plastic pitcher into a Dixie Cup.

"Drink up," he says, handing it to me.

"No, really, I—"

"No, really. If you're not careful, I'm going to have to put *you* in a bed, and we're running short on beds here."

I take a few sips of orange juice, and look at the doctor again. He's the kind of guy I should be going out with instead of Lenny. A nice young Jewish doctor.

"Tell me the truth." I finally find my voice.

He pauses for a fraction of a second, as if debating how much to tell me.

"The truth," he says. "Okay. The truth is, we just don't know."

"About my father? Or my mother?"

"Well, your mother has over eighty fractures, and she's lost a lot of blood, but the good news is, she doesn't seem to have any internal injuries."

"What are her chances?" I ask. I want the world divided into numbers, percentages.

"I can't make any guarantees. I'd say last night she had a thirty percent chance of survival. Now it's up to fifty-fifty."

I don't try to digest this. Instead, I shift gears.

"And my father?"

His eyes dart to the exit sign. He doesn't want to be having this conversation.

"Your father is an extremely complicated medical case," he says. "We're trying to get to the bottom of it."

I feel like telling him that my father is not a medical case, that he is a stockbroker, an observant Jew, a philanthropist, that he likes the Yankees and once had aspirations to play professional baseball. That he is loved by the woman with eighty broken bones, by the divorced shrink on her way here from New York City, and by the fucked-up girl with a face rash standing in front of him. But I say nothing of the kind. Instead, I ask him how I can help.

"We need to know what medication your father has been taking," he says. "Your—half sister, is it?—suggested that perhaps there might be"—here he pauses delicately—"quite a few possibilities."

"I can go to their house and empty out his medicine cabinet," I say. "I'll go tomorrow."

"Ms. Shapiro?"

"Yes?"

"Go today."

When I was thirteen, my mother discovered a lump—a swollen gland—on the right side of my neck. She rushed me to my pediatrician (I did not stop seeing a pediatrician until I went to college), and he fingered the gland, frowning.

"I think we should get this checked out further," he said.

I don't know exactly how it came to be that my mother, Roz, and I traveled from New Jersey to Boston Children's Hospital, where I went through three days of tests in an effort to get to the bottom of what caused the swollen gland. I'm not certain why no voice of reason spoke up and suggested that perhaps we should wait a few days just to see if the the gland settled down by itself. After all, maybe it was something simple, something less than lethal—like an allergy to shampoo.

In Boston, I had blood drawn; everything looked normal. I had X rays, and the side of my neck was poked and prodded by a team of residents. I had a spinal tap. Still, nothing. I waited in white hallways next to other children, emaciated children with yellow eyes and skin, while my mother conferred with the doctors.

Finally, they told my mother that the only way to be absolutely certain that nothing was wrong with me would be to remove the gland. The following morning, my head was wrapped in a turban (they had wanted to shave my hair but I threatened to run away) and I was put under while they surgically excised the gland from the side of my neck.

The following week, back in Hillside, I was doing my homework at the kitchen table when the telephone rang. We had two telephones side by side in the kitchen, a white phone and a black phone. The black phone number was given out only in cases of emergency. It was the black phone ringing. My mother leaped to answer it.

"Hello?"

A pause.

I looked up from my homework. Her face was trembling.

"Oh, thank God," she breathed.

She listened for another moment, nodding.

I fingered the thick bandage on the side of my neck.

She hung up and turned to me, her eyes huge and wet.

"I was so frightened, darling," she said. "But everything's all right now."

Until that time, I had not allowed myself to think that something could really be wrong with me. I was thirteen, and I had a child's sense of invincibility that amounts to a kind of faith. With each needle stuck in my arm, each doctor kneading the lump, I saw my mother's eyes sharpen with terror and that faith slipped away. By the time I returned home from that Boston hospital, I had absorbed my parents' fears and made them my own. We skulked like ghosts inside the safe, thick walls of the house in New Jersey, protected by emergency phones and alarm systems, tranquilizers and special air filters. Three of us, waiting for death.

The driver slows to a stop in front of my parents' new home and I stare at it through the car window. They moved here only a few months ago, and though I had meant to visit, this is the first time I'm actually seeing it. It's blander than I had imagined, and seems larger than the house in Hillside. Large and empty. No one has been here since the snowfall, and the driveway is un-negotiable, a smooth, untrammeled slope of white.

I am soaked from toe to midcalf by the time I get to the front door. Grasping a ring of keys—my mother's—I try to keep in mind a complicated set of instructions having to do with alarm systems and double locks.

The house I grew up in was protected by three different kinds of alarm systems: A steady red light outside the front and back doors switched on by a small circular key. A motion detector in specific, supposedly crime-prone rooms, activated by currents in the air. And carefully placed "panic buttons"—one in my parents' bedroom and another in the kitchen. When pushed, they set off an earsplitting siren in the house to alert the local police to a crime in progress.

Now, grappling with secret codes and keys makes me feel at home. I remember my mother's drugged, mumbled suggestion that it might be easiest to enter through the garage door, so I make my way around the side of the house, my feet numb in my boots. I turn the key on the side of the door and it rises with a groan. Inside, there is one car, my father's Subaru. His tweed cap is hanging on a hook by the inside door, and his galoshes are on the floor.

It is Saturday. My parents have not been home since some-time on Thursday, and the kitchen—the first room I walk through—still has life in it: a half-empty coffee cup next to the sink, Thursday's *New York Times* folded open to the national news, a blinking light on the answering machine, a shopping list written in my mother's pointy script. *Eggs, Special K, 2% milk, tomatoes, salad stuff, boneless chicken breasts*. Catalogs and bills are piled on the kitchen table, and at the center of the table, just where it's always been, is the lazy Susan with my father's drugs.

I'm a woman with a mission. I am so far beyond my saturation point that my mind has actually become sharp, methodical. I look under the kitchen sink, where I find my mother's stash of paper bags. They range from Food Emporium to Giorgio Armani. My mother keeps everything, and I keep nothing. I pull out the largest bag—lavender, from Bergdorf Goodman, as it happens—and begin to fill it with the prescription medication and vitamin bottles from the lazy Susan. Valium, codeine, Percodan, Percocet, Empirin, phenobarbital, papaya tablets, garlic pills, bee pollen. It all goes in the bag.

I climb the stairs of my parents' new house, feeling like an intruder. The art and furniture I grew up with are all reshuffled, and look familiar but slightly off, like seeing the next exposure of a well-known photograph. I hope I haven't forgotten to turn off any of the alarms. If I trip off a siren, if the police come, what will I say? That my parents are in intensive care, and I'm here to retrieve my father's drugs? There are pictures of me all over this house: baby pictures, one or two from high school, and a recent eight-by-ten that I use for commercial auditions. Will that be enough to prove that I belong here?

Inside my parents' medicine cabinet there are more amber prescription bottles. I begin trying to sort them out, but end up just dumping the whole mess into the shopping bag. They are almost all painkillers. I'm tempted to reach in and grab a few, but I've never been one for pills. Give me scotch or cocaine anytime, but pills make me feel out of control.

The bag rattles as I move through the master bedroom. There is a large picture window overlooking the swimming pool, bare-branched trees skeletal against a steel-gray sky. The configuration of my parents' bed has not changed; two twins pushed together, according to the Orthodox custom that dictates that husbands may not sleep with their wives when they're menstruating. Hasn't my mother hit menopause? Wouldn't that render the point moot?

I pass the framed eight-by-ten once more. In it, my hair is tousled in an eighties shag, and I am smiling a smile fit to sell toothpaste. My skin is airbrushed, the whites of my eyes doctored to look brighter. In the bottom right-hand corner, almost obscured by the frame, my stage name is printed in bold black letters: Dani York. According to my agent, "Shapiro" just doesn't cut it in the world of blond, blue-eyed California girls. I arrived at my stage name while on vacation in Maine with Lenny. We went through towns in Maine: Dani Portland, Dani Kennebunkport, Dani Ogunquit, until we arrived at York.

On the first floor, I rest for a moment in the mustard-colored chair where my father used to sit, his neck in traction, before his spinal surgery. I close my eyes, remembering the way the contraption wrapped around his jaw and cheeks, squishing the loose skin together until he resembled one of those wrinkled Chinese dogs. He would sit there for hours watching *Hogan's Heroes* and *Gilligan's Island* reruns, a thick cord extending from the top of his head to a device on top of the doorframe to relieve the pressure on his spine.

Next to the chair, on an old walnut end table, there is a telephone and a pile of *Prevention* magazines. *Prevention* may well be my father's favorite publication, a hypochondriac's version of *Playboy*. Eat garlic tablets and you won't have a coronary! Papaya enzymes prevent ulcers! How to raise your "good" cholesterol!

What did my father suppose were the health benefits of a shopping bag filled to the brim with a stunning array of half-empty bottles of tranquilizers and painkillers?

I dial Lenny's office in the city and get his voice mail. I check my watch and realize it's lunchtime. Lenny's probably at Smith & Wollensky's, around the corner from his office, digging into a juicy porterhouse. He's not worried about his cholesterol. Lenny is twenty pounds overweight, smokes fat Cuban cigars, drinks fine red wine and vintage brandy. He doesn't go to temple on Rosh Hashanah and Yom Kippur and pray to be forgiven for his sins. I'll bet he's never even heard of *Prevention*, much less believed in it as a concept.

"Lenny, it's me," I say. "I'm back. I'm at my parents' house— oh, never mind, you can't reach me. I'll try you later."

I hang up the phone and stare at the wall. It's a perfect wall, without a smudge or a missing chip of paint. An illuminated Hebrew manuscript page hangs in a gold frame, and below it there is a small, colorful weaving. On a leather chair across the room, there is a needlepoint pillow my mother made when she started her own tennis product business, of a yellow ball with a blue eye at its center. Underneath it, the motto reads *Keep your eye on the ball*.

I pick up the phone and dial Lenny again.

"It's me again," I say into his voice mail. "I need to see you tonight. This is a mess, this thing with my parents. A horrible, total, fucking mess."

Of course, seeing Lenny tonight is the last thing I ought to be doing. If I go back to the city at all, it should be to a cup of warm tea and the comfort of my own bed. But I'm terrified to be alone. *A horrible, total, fucking mess.* Well, it really is, isn't it? So why do I feel that I'm lying? That I'm on some sort of drug trip, a hallucination? That I'm making all this up? I've been lying to myself for so long now that I don't know what the truth might feel like. But now, ready or not, the truth is all around me. I hear it in the wind chimes clanging in the bitter February wind, the plastic bottles rattling like bones.

I double-lock the back door to my parents' house, punching in the alarm codes, making sure it's safe and sound. I touch my fingers to the silver mezuzah on the outside doorframe, then lift my fingers to my lips: *Bless this house.*

CHAPTER
THREE

I am fourteen and my parents are fighting all the time. I hear their voices rise and fall behind their closed bedroom door. My mother's voice is more insistent than my father's. I can't make out what she's saying—only that she's very angry—and I can barely hear my father at all, except for the occasional booms of rage, loud outbursts after which everything is quiet.

I'm afraid for my father. He's just returned home from a month-long stay in the hospital, where he had a spinal-fusion operation. The surgery has left him temporarily paralyzed down the right side of his body, so now he spends his time alternately in traction and doing his exercises to regain the use of his arm. He creeps his fingers up a wall, propping his useless elbow with his good hand, and his face is so haggard and angry that I'm afraid to go near him. He's taking more pills than ever now—twenty-five milligrams of Valium at a time.

One night, while I'm in my bedroom doing homework, I hear my mother shriek. She is shouting my father's name over and over—*Paul! Wake up, Paul!*—and I bolt down the hallway. My father is lying on the pink tile of the master bathroom floor, his eyes open, pupils fixed on the ceiling. His bare feet are sticking straight up from the bottom of his robe. His face is the color of chalk. My mother is crouched over him, and when she sees me she screams to have me call Dr. Kogan. Kogan is a gastroenterologist who lives next door.

I race down to the kitchen, find my mother's phone book, and call Dr. Kogan. My heart is pounding. I think of the way I have seen my father press two fingers into the side of his neck and look at his watch, counting his pulse. I close my eyes tight and pray. *Sh'ma Yisroel Adonai Elohenu, Adonai Echad.* I don't go to the yeshiva anymore, and I've started to sneak around and eat bacon and go out on dates with non-Jewish boys, but I'm hoping God will understand that the prayer is for my father, not for me.

Dr. Kogan rings our doorbell two minutes later. He's wearing sweatclothes and carrying a black bag. He follows me upstairs, through my parents' bedroom and into the bathroom. My father is sitting up, hunched over on the bathroom tiles. His glasses are askew.

"Hi, Eddie," he says to Kogan weakly. "What's up?"

"I should be asking you that, buddy," says Kogan.

My father smiles sheepishly. His pupils are like pinpricks floating in his pale green eyes.

"Got a little groggy, that's all. Tripped and fell."

I'm on my way to meet Lenny for dinner at an Italian restaurant on the Upper East Side. I have been awake for thirty-six hours. I should go home and take a bath, maybe ask Lenny to come over and give me a back rub until I fall asleep, but conserving my strength is not part of my repertoire. Instead, as the car idles in bottlenecked traffic at the mouth of the Holland Tunnel, I call my friend who happens to be a cocaine dealer and ask if I can stop by her loft on my way uptown.

Not doing coke has been one of my recent promises to myself. In the beginning of my 1986 journal, I have written this list of New Year's resolutions:

1. Start eating healthy
2. Cut down on drinking
3. No c.!
4. Get back to acting class
5. College courses?

Thus far, I've managed to make a dent in only one resolution, by calling my old acting coach, Fred Kareman, and asking him if I can come back to class. Freddy is considered one of the great acting teachers in the city. I'll never understand why he let me into his class in the first place. The competition for each slot is fierce; Freddy handpicks his class out of dozens of actors and actresses, and expects a serious commitment to "the work"—the Sanford Meisner technique—in return.

But I find it impossible to be a student of acting and the mistress of Lenny Klein. When Lenny asks me to travel with him, which has been at least once a month, I never say no. Being Lenny's girl is my full-time job, and everything else is a hobby. I have recently resolved to change this. But I have also sworn up and down that I wouldn't do cocaine anymore, and here I am, pulling up in front of a small loft building on Wooster Street and telling the driver I'll only be a minute.

I cannot face dinner with Lenny without a vial of cocaine in my handbag. My friend buzzes me in, and I run the five flights up to her loft, skimming my hand along the chipped wood banister. I ring her doorbell, and she opens it, holding a paper bag. I know what will be inside without even looking, and I hand her a check for one hundred dollars. She trusts me and always takes my checks—a nifty trait in a drug dealer. Behind her, a mirrored wall of closets is shattered by a single bullet hole; cracks in the mirror splay out from the center like crude rays of sunlight.

Lenny is sitting in a corner banquette in the back room of a dimly lit Italian restaurant. As I follow the maître'd past tables covered with stiff white cloths, I hear his laugh rise above the dull clatter, and see that he is not alone. A gray-haired man and a young woman are with him. The woman has dark hair, and for a brief, insane moment I think it's Jess, until I get a little closer and see that she's in her early thirties, a decade older than Jess or me.

I haven't actually spoken with Lenny yet. He left me a message with a place and time to meet—but he didn't say anything about people joining us. Tonight is hardly a good night to make new friends, and I feel a flash of anger, almost enough to make me spin on my heel and walk out of the restaurant. But where would I go? I don't want to be alone. The vial of cocaine in my bag and the chilled bottle of Chardonnay on the table will get me through the night. And besides, Lenny will have a reason. Later, he'll tell me he thought it would be a good distraction, or that these people are important out-of-town clients. Whatever. He grins when he spots me, half rises from behind the banquette, says, "There's the Fox!"

He kisses me full on the lips, lingering too long, then introduces me to our dinner companions. As I shake their hands, I recognize the gray-haired man as one of Lenny's partners in the law firm, the guy who handles all the Hollywood business. I've met him before. I don't know anything about the woman he's with, except I'd lay odds that she isn't his wife. I've never met the wives of any of Lenny's friends or partners—but I have met the girlfriends. After almost three years, I'm beginning to get it: these men have parallel social lives that never intersect.

"Dani's had a hard day," says Lenny.

He flags the waiter, and I order a vodka martini, straight up with two olives.

"What happened?" asks the brunette. She is wearing a large, round diamond dangling from a simple gold chain. It's the kind of necklace no one who rides the subway would ever wear.

"My parents were in a car accident," I explain woodenly. I'm not feeling anything, except maybe that vague, weird sense I had earlier in the day, of lying through my teeth. Why else in the world would I be sitting with total strangers in a loud trendy restaurant? My relationship with Lenny is based on lies. I wonder, is there any chance he thinks I've made this up? I suppose, in the world according to Lenny, a lie of this magnitude would not be out of the question. I notice him watching me carefully, his eyes steady over his drink. Lenny prides himself on being a good judge of character. He is a trial lawyer, after all, and penetrating the defense is what he does best.

"So, how are they doing?" he asks.

I tell myself he's concerned, that he really cares about me and my family.

"Not so great," I say. "They don't know what's wrong with my father—whether it was a stroke, or his head—"

"I'll call Burt Zuckerman," says Lenny. "He's the best neurologist in town. He'll arrange to have your dad transferred to Mount Sinai."

"He can't be moved."

Lenny shakes his head, as if he knows better.

"Leave it to me. Burt and I are old buddies—our kids went to nursery school together."

"No, Lenny. I mean, the doctors say it's dangerous to move him."

Lenny smiles at me condescendingly. He always knows better. He has convinced me that the real world operates on a certain level—above my head—and that I will be able to survive in this world only if I stick with him. My parents, according to Lenny, are as vulnerable as toddlers, with their religious beliefs and quaint customs. They have not prepared me for the wilderness out there, a dark place where men have wives and mistresses, where politics and friendship are indistinguishable, and money smooths it over like an analgesic.

I excuse myself.

There is a phone near the ladies' room. I fish from my bag the matchbook on which I've scribbled the number of Overlook Hospital and call the ICU. The nurse who answers tells me that my father's condition is still listed as critical, and my mother has been bumped down to serious. There are no phones by the beds in the ICU, so my mother has asked the nurse to be sure to tell me if I called that it's going to be all right.

I wonder if my mother would think it's going to be all right if she saw my father lying on the opposite side of the unit, breathing through an oxygen mask, his eyes closed and sunken. I wonder if she'd think it's all right that her daughter is shutting herself into a ladies'-room stall and unscrewing the small brown glass bottle half filled with cocaine and tapping a bit into the concave spoon of her pinkie nail and snorting it quickly up her nose.

I sit on the toilet, fully clothed, and snort a few more pinkie nails' worth—probably about an eighth of a gram. I want to save the rest for later. Later, Lenny and I will go back to my apartment. We'll stumble into bed and claw at each other; he'll whisper lies to me, and I'll believe him.

I'm not alone in the ladies' room. I flush the toilet just for show, then open the stall. The brunette is standing by the sink, reapplying her lipstick. She has tiny lines around her mouth.

"Hey there," I say weakly.

I splash cold water on my face, checking my nostrils in the mirror for caked white powder. Out of the corner of my eye, I see her look at me from head to toe.

"Lenny's cute," she says.

I try to smile, but my jaw feels wired shut from the coke, my eyes pinned open. As we leave the bathroom I notice that someone has carved a heart on the door, just above the knob. But there's nothing inside it. Only empty space, waiting to be filled in.

Here, in no particular order, are some lies Lenny has told me: that he and his wife don't sleep in the same bed; that they haven't had a "real marriage" in years; that she is mentally ill; that she is undergoing electroshock treatment in a clinic outside Philadelphia; that he has cancer, and has to fly to Houston three days a week for chemotherapy; that his youngest daughter, age three, has a rare form of childhood leukemia. Lenny cannot get a divorce for all of the above reasons. He is heartbroken that he cannot leave his wife and marry me.

For a long time I believed him. With every bone in my body, I trusted that Lenny Klein was telling me the truth. After all, when we talked about it, his jaw tightened, and his big brown eyes filled with tears. His voice quavered with pent-up, complex feelings I couldn't possibly begin to understand. Poor Lenny! I marveled that so many bad things could happen to one person, and I vowed to take care of him. Writing late at night in my extensive journals, I exhorted myself to be a real woman—one who could step up to the plate and be good to her man in his moment of crisis.

Years from now, I will hold Lenny's lies up to the sunlight and examine my own motivations for believing what, in retrospect, seems preposterous. I will reread my old journals and notice the way my girlish handwriting deteriorated into a scrawl as I wrote *I have to be there for Lenny. He needs me,* and *He's going through so much. I don't know if I can handle it—but I have to be strong!* I will try to remember that Lenny was a trial lawyer, that he built an international reputation based on his own pathology: he lied with an almost evangelical conviction. He prided himself on being able to convince anyone of anything.

I will remember Paris, 1985. We are walking along the Boulevard Saint-Germain on a cloudless spring day. The rooftops of the Left Bank are creamy against a rare blue sky, and the air outside Café de Flore smells of croissants and the acrid smoke of Gitanes, but I don't notice. It is only years later, as a grown woman, that I will take in the rooftops of Paris, the extraordinary sky, and so I am supplying this scene with a collage of my own memory. In Paris, 1985, I see only what is within one square foot of me, too busy feeling the complicated stew of sensations being with Lenny provokes. I am hungover, floating on a wave of last night's Puligny-Montrachet and a four-star dinner that wound up in the toilet of the Hôtel Ritz. Lenny's arm is around me, thick and proprietary, and it reminds me of the sex we had that morning, the way he pinned me to the bed and didn't let me move my arms until I came hard, fast, in spite of myself. In Paris, I am like an animal curled in a patch of sunlight, interested only in the beating of my own heart. Sex, wine, food, sleep. I am a physical being, living on the other side of a clear, thin membrane that separates me from anything having to do with the world.

I have not read a newspaper or spoken to a soul other than Lenny for weeks now. We have been to London, Monte Carlo, the Côte d'Azur. I have played blackjack in private clubs with oil sheikhs who asked me to blow on their dice for good luck; I have driven a convertible around the hairpin turns of the Moyenne Corniche; I have eaten langoustine on a boat floating somewhere off the shores of Cap d'Antibes. I wear dark glasses and haute couture suits, a gold watch, and a long, thick strand of pearls. I have no idea who I am.

Lenny steers us onto a narrow side street off the Boulevard Saint-Germain, and into a children's clothing store filled with the embroidered little dresses my mother used to buy me as a child. He tells me he wants to buy a dress for his youngest daughter, the one with the rare form of childhood leukemia. I help him look through racks of tiny dresses suitable for a three-year-old, until we find one he deems perfect, a pale yellow silk smock with a Peter Pan collar. He holds it up to the sunlight, and his eyes fill with tears. *She'll never live to grow out of this dress*, he whispers. *My baby girl*.

By the time my parents are in the Overlook Hospital ICU, I have known the truth about Lenny for quite a while. It takes one to know one, and I have so little regard for the truth in my own life that I have developed a sixth sense, radar for his deceit. He has layered his lies one on top of another until they have become opaque, an elaborate construction resembling reality. He is fond of quoting probably the only line he knows from Franz Kafka: *White is black and black is white*, he often says with a sigh. I don't know exactly what he means by this, but it seems to have a lot to do with my current life.

It started small, months ago: he called me from a business trip, told me he was in the Montreal airport to catch a flight to Calgary. I checked with the airline and found out that the flight should take approximately five hours. So when Lenny called an hour later to say he had landed in Calgary, I very calmly asked him where he *really* was.

Calgary, he said.

No, Lenny, *really*.

He stuck to his story. In the time that I knew him, he never, ever, changed his story midstream. I hung up on him and called his family's house in Westchester. When the maid answered the phone, I asked to speak with Mr. Klein. And when he picked up the extension and I heard his rough, craggy *Hello?* I screamed so hard into his ear that he dropped the receiver.

He raced into the city. He let himself into my apartment and found me curled up in bed. He scooped me up and held me to his chest. His wife wasn't home, he told me. She was having a shock treatment. And someone had to take care of his daughter. He hadn't wanted to tell me because he wanted to spare me, to protect me from the horror of his life. Surely I understood. *Ssshh, sweetheart,* he murmured into the top of my head as I wept, my face beet-red like a little girl's. *So many people need me, but I love you best of all.*

After dinner with Lenny's partner and his babe du jour, we go back to my apartment. We are in bed fucking when the phone rings. I move my hand to answer it, but Lenny grabs my wrist. I think it might be the hospital calling about my parents, but before I can articulate the thought, the machine picks up.

"Dan, it's Susie. Are you there?"

I reach across my nightstand to the phone. The digital clock reads 2:37 A.M. Lenny is on top of me, deadweight, smelling like Speed Stick and traces of English Leather. I try to say hello into the receiver and realize that I'm dead drunk. The cocaine is wearing off, and the room is spinning.

"The alarm has gone off in Tewksbury," says Susie. "The police called me."

"Yeah, so is it a break-in?" I mumbled. I imagine some resourceful criminal finding out who's in the hospital and burglarizing their empty house. What got set off? Radar, motion detector, panic buttons, hidden pads?

"No, just some faulty wiring, I guess. You were there this afternoon, weren't you?"

God, was it only this afternoon? What is she saying? Did I screw up the locks somehow?

"Uh-huh."

"I just thought you should know," she says. "They've fixed the problem for now, but I gave them your number in case anything more goes wrong. I'm whipped, and I have a seven-fifteen patient."

Implicit in this is that I have all the time in the world, which, in fact, I do. Susie does something with her life, actually helps sick people. She's busy being a shrink, and I'm busy needing one.

"Susie, is everything going to be okay?" I ask her. Lenny is moving his tongue down my side, as if we're on some sort of vacation. He moves his head between my legs, but I feel nothing.

"I have no idea."

I push Lenny away, and he groans loud enough for Susie to hear him.

"God, Dani, is that asshole there with you?"

I don't answer.

My half sister says good night and hangs up the phone. I imagine her turning on her side, alone in her big bed in Greenwich Village, safer and less alone than I am with Lenny's arm flung over me. He is snoring.

The accident has supplied me with something I had been missing until now: structure. Before, my life revolved around auditions and Lenny. Sometimes I had nothing to do, and I sat in my darkened bedroom in my light blue terry-cloth bathrobe, smoking cigarettes and watching soaps, telling myself I was checking out the competition. After all, I auditioned for soap operas regularly: *All My Children, One Life to Live, The Edge of Night*. I had been seen for every young female role on the ABC lineup, from angelic blonde to vixen from hell. On those down days I would turn off the ringer on my phone, adjust the volume on my answering machine, and shut out the world. I would move only from my rocking chair in the corner of my bedroom to the bathroom, where I would pee, weigh myself, then stare into the mirror.

But when I had calls for soaps or commercials, I had to pull myself together. I would curl my hair in hot rollers and carefully make up my face, using special products meant for the camera that looked heavy and cakey in broad daylight. I would make my way to casting directors' offices or midtown advertising agencies, where I would wait in a brightly lit hallway with a half-dozen other girls who looked vaguely like me. I would read the xeroxed copy, looking for ways to recite such sentences as "Diet Seven-Up is the one" with my own unique inflection.

Evenings, I would meet Lenny at a bar near his office where we would have a few scotches while deciding what to do for dinner. Often dinner was only a concept, something to give shape and direction to the evening. Certainly, I didn't want to eat. Eating meant finding a bathroom and making myself throw up. And all Lenny really wanted to do was go somewhere where we could rip our clothes off. He didn't care if that place was a $300-a-night hotel room or my apartment. In fact, he probably preferred hotel rooms, with their faint whiff of sex for money.

Did I think this was romantic? A man and a girl in a bubble? Thwarted love, perhaps, or the pure, clean lines of obsessive lust? Time has a way of building a frame around those moments, leaving me not with a mental picture of how I felt back then, but only with a sense memory: the cold blast of air-conditioning on a humid August night, my legs scissored apart by Lenny's thick hands, the sour slosh of white wine in my belly and my lips so numb from cocaine they don't feel a thing. I am reduced to body parts: breasts, hips, stomach, cunt. Lenny tells me women are over the hill at thirty. It seems an age I will never reach.

Now, in my new routinized life, I have a purpose, a reason to wake up in the morning. I am needed. The centerpiece of each day is a medical crisis. Yesterday my father had an operation to insert a pacemaker. The day before yesterday, my mother developed an infection near the metal pin that runs through her shin and needed a heavy dosage of antibiotics. I have learned the language of good, stable, serious, critical, and how it applies.

Today, the fifth day, my father wakes up. When I walk into the ICU, I see him sitting up in bed, his glasses perched on his nose. My heart leaps, and my body follows. I am by his side in a flash, my arms wrapped around his neck, my lips pressed against his hot, stubbled cheek. *Thank you, thank you, thank you,* I whisper, smelling him, breathing him in. For the past several nights, before I've gone to sleep I've run through every prayer I know by heart, thinking that if my father can't pray for himself, perhaps mine will do. For the first time since I received the phone call, I allow myself to imagine that maybe, just maybe, the God my father has believed in all his life will come through for him.

"Hi," my father says weakly. His voice is the most beautiful sound I've ever heard. The minute I hear it, I realize I had thought I'd never hear it again.

"Hi, Daddy."

I look at his face, sallow and sunken, his pale green eyes

magnified by his wireless spectacles. The planes of his cheeks, the slope of his forehead, his bulbous nose, are my map of the world. When I was a little girl, he used to hug me too hard. He would clasp me in a bear hug and practically crush me in his grip, as if he could fold me into him and never let me go. When he'd finally release me, I felt his love in my aching bones. I would give anything for one of those hugs now.

He smiles at me beatifically.

"Where am I?" he asks, and in three small words it all collapses around me.

"Where do you think you are?" I ask.

My senses sharpen. I now see that my father's eyes are kaleidoscopic, like a cartoon character's. I hear the steady beeps of his EKG, and smell the antiseptic they use here, stronger than whatever it kills. I taste metal on my tongue, and touch my father's forehead, a futile, maternal gesture, as if he were a toddler with a fever and I'm taking his temperature.

"Stock exchange?" he asks, his eyes brightening.

"No."

"Hotel?"

"No."

"I give up."

My mind races with what to do. Where's Susie? She's the psychologist. She's also the adult. I think of my mother, just across the semicircle of the ICU. Does love make her psychic? Does she know that her husband and daughter are only fifty feet away?

"You're in the hospital, Dad," I say.

He stares at me blankly.

"No."

His voice has the inflection of a five-year-old about to throw a temper tantrum. Behind his eyes I see something, a flash of terror. Somewhere inside him, in some pocket untouched by the mangled, fluid-soaked part of his brain, he knows.

I grab his hand.

"You were in an accident," I say softly, stroking his fingers.

"No!"

He looks at me wildly, shaking his head violently back and forth. His jaw is slack with fear. His mouth twists into a grimace.

"Where is my wife?" he croaks.

"Mom is here," I answer.

"Where?"

"She . . . had to go out for a little while."

He glares at me as if I'm the enemy.

"I don't believe you," he says flatly.

Never in my life have I heard my father speak this way to me, or to anyone. It is as if all propriety, all gentleness has drained from him. He is a boy trapped in a sixty-four-year-old man's body. I think of the sepia photographs of my father as a toddler, the round, innocently cruel, happy eyes of childhood emerging, ghostlike, in his green, haggard face.

"I want to see my wife!"

He tries to swing his legs over the side of his bed. I keep a grip on his hand, imagining that I will somehow be able to arm-wrestle my two-hundred-pound father and keep him in bed if I need to—but he falls back onto the pillows, sapped. He has been in a coma for five days.

The nurse comes scurrying over to us, her white rubber-soled shoes squeaking against the linoleum. She's one I haven't seen before, and I find myself unreasonably mad at her, as if she somehow should have been able to prevent this.

"Now, Mr. Shapiro, you're not being a very good boy," she says in a singsong, pulling the thin white sheet over his hospital gown, up to his chin.

I look at her, stupefied. My father is an elegant man, and an imposing one. He wears navy-blue suits and silk ties, shiny black shoes and cashmere socks. He has his own seat on the New York

Stock Exchange. His Wall Street colleagues have long referred to him as "rabbi" because he takes off Jewish holidays they've never even heard of. He has spent his life blending worlds: Orthodoxy and high finance, religion and culture. And now, a nurse is telling him he's not being a very good boy.

I take her aside.

"Please don't talk to my father that way," I say, trying to smile, feeling my lips tremble.

She looks at me with what appears to be pity.

"It makes no difference to *him*," she says.

I look over at my father. He is holding his hand in front of his face, curling his fingers, staring at his cuticles with intense fascination. He has pushed off the sheet once again, hairy legs exposed. I have never seen my father naked, and don't want to start now. He catches my eye and winks. It is an entirely uncharacteristic gesture; my father just isn't a playful guy. I feel a wave of anger at him. *Snap out of it!* I want to scream.

I go down to the hospital coffee shop and try to gather my thoughts. What am I going to tell my mother? She has wanted the truth all along, and I've given it to her. When she's asked me if my father is "awake yet"—a euphemism for asking if he's still in a coma—I've answered her truthfully. But now the rules have changed. My father is awake, all right, but he has woken into someone else. I guess my mother, Susie, and I all believed that he'd either stay in a coma and slip away from us, or that he'd miraculously come to and be the same man who was behind the wheel of that Audi 5000 in the moments before he passed out. This child-man, this stranger in my father's body, may be more than my mother can bear.

I drink two cups of dreggy coffee and force down a piece of whole wheat toast. I flip through the first few pages of the Metro section of the *Times*. Lenny is litigating a high-profile case, and sometimes there are updates in the news, along with photographs of him leaving the courtroom in his pin-striped suit and trademark raccoon coat. I flip through the paper looking for his picture, thinking it might bring me some comfort. Somewhere along the way I have grown to believe that comfort is to be found along the blade of a knife. What hurts me also stops me from thinking, from feeling. So when I see the photo of Lenny on the courthouse steps, when I am struck for the millionth time by the fact that I occupy a small, invisible corner of his life, everything around me grows a bit dim: the edge of the counter, the paper place mats with their local guide to Summit, the Muzak version of "Norwegian Wood" piped into the coffee shop.

I chew the toast slowly, forcing myself to swallow. I even take a few sips of orange juice. In the past five days since the accident, it has begun to occur to me that I must conserve my strength. My father has Susie and me, but my mother has only me. If I fall apart, so will she. My mother told me that in the split second before the car hit the divider, she saw my face and was filled with an indescribable sadness. *I couldn't bear leaving you alone in the world,* she whispered, her eyes black-and-blue. *So I hung on.* And now it is I who cannot bear to leave my mother alone in her little curtained world for even a moment, her husband lost to her, her stepdaughter filled with animosity. I have no choice. After a lifetime of rebelling against her, of trying so hard to separate I ripped myself up in the process, now it seems we are attached at the hip. I am her lifeline, supplier of love, support, and information. It's not a role I could have imagined, but now that it's here, it seems it could have turned out no other way.

"My beautiful daughter," she pronounces from her pillow when I part the curtains and poke my head in, half hoping she'll be asleep. My mother looks worse each day, which the doctors assure me is to be expected. Her face is less swollen, but more deeply bruised. Everything is slightly askew, a Picasso portrait, nose and eyes jarred out of alignment.

"Let me look at you," she says.

I walk to the side of the bed and crouch so we're at eye level. Her gaze darts eagerly over my face. My mother has always loved to look at me, and I have never gotten used to it. I squirm under her examination. She often says things like *Look at that perfect profile*, as she turns my head from side to side, fingers cupping my chin. Or she'll utter the only nonliturgical Hebrew phrase she knows: *Eza bat yesh lee*, which roughly translates into "What a daughter I have." She means well. I'm sure she had no intention of raising me to crave this kind of attention, to the point where I feel something's wrong if it's not there.

"Are you a little tired?" she asks. "You have rings under your eyes."

She must be getting better.

"You're not exactly looking rested yourself."

"Well, I must be doing okay, because they're moving me to another floor," she says.

"Mom, that's great!"

I assume it's great. After all, there isn't anywhere to go but up from the ICU.

"I asked them to see if they can put your father nearby."

She looks at me carefully as she says this. I'm sure she wouldn't put it past me to lie to her about how my father is doing. For all she knows, he might not even be alive, though I don't think she has allowed herself to consider that possibility. She has suspended her disbelief, buoyed by the almost manic optimism with which she has gotten through life. How much should I tell her? Her condition is fragile, her emotional state in need of protection. With eighty broken bones, a broken heart is out of the question.

"Daddy's doing a little bit better," I say. It's not exactly a lie. What could be worse than a coma?

Her eyes fill with tears.

"See? I knew everything would be all right," she says. She reaches for my hand, the same hand my father held only an hour ago.

"I want to see him," she says.

My mind races. I wish there were someone with me right now—Susie, or Morton, anybody to help me out with what to say or do.

"I'm not so sure that's a good idea—"

"Why not?"

"It might upset him to see you like this," I say, solving one problem and creating another. My mother has not yet seen herself in a mirror.

"My purse is over there," she points her chin at a small metal dresser, "in the top drawer."

"Mom, I—"

"Bring me my compact."

She says this in her emphatic way that brooks no dissent. One of my mother's favorite expressions is a line she misquotes from Plutarch: *Caesar's wife is above reproach*, she is fond of saying when she feels in some way maligned. I've always hated that. I mean, what does Caesar have to do with a New Jersey housewife? And anyway, does that make my father a Roman emperor? My mother grew up on a chicken farm in southern New Jersey. Her father was a Russian immigrant. She went to college on a full scholarship, married and divorced an assistant genetics professor before she was thirty. But her favorite part of her history—the part she tells me again and again—has to do

with her dating life before she married my father: the Chanel
heir who drove a white Cadillac by day and a black Cadillac by
night; the count she met on her first trip to Europe; the magnifi-
cent clothes her own mother sewed for her, straight from the
pages of *Vogue*.

She locks eyes with me, waiting to see if I'll do her bidding.
After all, this is the first time in my life I've had power over my
mother. Her hospital bed has become a crib, and I am the
looming adult who can grant her wishes—or not. I am horrified
by the responsibility. I can barely manage to pay my bills and
keep fresh milk in my refrigerator.

"Mom, you don't look so good," I say gently. "Why not wait
until—"

"I have a right to see my own face," she says.

Behind the icy bravado there is terror. It lurks just beneath
her skin. Maybe whatever she is imagining is even worse than
what's actually there. I open the metal drawer of the dresser and
pull out her handbag. It's an old leather purse, black and quilted,
and I remember it from my childhood. This must be what she
was carrying the night of the crash. One of the few remnants of
my mother's chicken-farm roots is an inability to throw anything
away. She keeps her cashmere sweater sets from college, her
boots that have been resoled a half-dozen times, papers and clippings that have long since lost whatever relevance they may
have once had. The cabinet beneath her bathroom sink at home
was always stacked high with free department-store samples and
cosmetics-company giveaways: hundreds of creams, lipsticks,
mascaras—more than could ever be used in a lifetime—lined up
next to dozens of unopened boxes of perfume bought in airport
duty-free shops.

I rummage through her purse for a compact. Rummaging
through my mother's purse is nothing new to me. As a child, my

desire to find out the real truth (which I also defined as whatever I didn't know) began with my mother's handbag. The romance of little slips of paper, contents of wallets, to-do lists, lipsticks! At twenty-three, my mother's purse has been replaced by Lenny's briefcase. Whenever I get a chance, I look through his papers, bills, correspondence. I think I will be able to patch together a complete portrait of Lenny, if only I can find all the pieces.

My fingers light on the smooth, round case of my mother's compact. It's small. She will only be able to see her face a fragment at a time. I hand it to her, then stand just behind her shoulder as she opens it, so I can see what she sees. I stroke the top of her head, the layers of her brown, blond-streaked hair.

She doesn't make a sound as she moves the mirror slowly in a circle, taking it all in: black-and-blue eyes, broken nose, fractured cheekbone. When she is finished, she snaps the compact shut.

"Okay," she says thickly. "Not too bad."

A week after the accident, my agent calls. He sounds excited, but then again, he always sounds excited. Good agents are brokers of hope, believers that this audition, this callback, this screen test, might be the one. He tells me that the head of nighttime casting at NBC saw my eight-by-ten and wants to take a meeting.

"Just a meeting," Sheldon says. "You know the drill. Just go up to his office looking like the traffic-stopping babe that you are, and charm his pants off."

He doesn't mean this literally. Or does he? Sheldon has come around his desk more than once in the three years he's been my agent and given me hugs and kisses that lasted a little too long. He's a forty-five-year-old married guy who wears a gold pinkie ring and a mustache that usually billboards whatever he's eaten for lunch. Sheldon handles me for film and television, not commercials—which means he hasn't made much money off me so far. But he seems to believe in me, and he sends me up for everything: Broadway plays, feature films, sitcoms, soaps.

I've told Sheldon about my parents, which elicited an offer to come over and give me a neck rub. His office sent flowers to my apartment, with a note written in the florist's script: *Our prayers are with you, from all of us at Entertainment, Inc.* I guess he thinks a week is enough time for me to be back on my feet, ready to roll. What's more, I seem to think so too. I tell Sheldon I'll be at the meeting, no problem.

Sheldon schedules me for three o'clock. He usually manages to make my appointments after lunch. Does he know how hard it is for me to pull it together in the morning? That on days I have auditions, I have to wake up early, take at least one aerobics class, and spend an hour on my hair and makeup before I begin to look halfway decent? Years from now, when friends see snapshots from this period, they will ask me point-blank whether I've had plastic surgery. My face is bloated from drinking, puffed full of air, and my body is incongruously thin. My parts don't look as if they go together. Nothing fits. But somehow, on camera, it all sort of works. My big blond head fills the screen nicely, and I have bone structure that seems to come out when a video camera is pointed at me, contours emerging from beneath the bloat.

On the morning of the meeting at NBC, I skip my hospital visit, planning to go at the end of the day instead. It has been weeks since I've been on any kind of audition, and I'm nervous. I hate auditioning. Since I've started acting and modeling I've had to work on making my face as still and placid as the surface of a lake. This requires some effort. I remember, as a child, my mother telling me not to frown or my face might freeze.

My mother has often claimed that I could have had a big career as a child model, but she pulled me out of it because she didn't want me to turn into a messed-up kid. When I was nine months old I was the Beechnut Baby Food baby in the national commercials, and when I was two I was the Kodak Christmas poster child. My face was plastered all over the country, in advertisements, on billboards. When I asked her why I was doing it in the first place—it certainly wasn't my idea at the age of two—she's answered that it just sort of happened. Accidentally. I wonder what my father thought of his little yeshiva girl wishing the world a Merry Christmas. My mother claims he got a kick out of it.

Jittery, I throw on a leotard, my mink coat, and sneakers, then walk down Seventy-second Street to the aerobics studio where I spend most mornings, when I'm not too hungover, taking a two-hour aerobics class called Fat Busters. I pass Fine & Schapiro delicatessen on my way to the studio, and the smell of fresh corned beef brings my father into such a sharp focus that for a moment I think I see him in the window, ordering his favorite sandwich from the guy behind the counter. I stop in and order a sandwich to sneak into the hospital for my father. Turkey, tongue, coleslaw, and Russian dressing, with a sour pickle on the side, and a Dr. Brown's Cel-Ray soda. It's a meal only a Jew would order, and the deli guy looks at me slyly out of the corner of his eye as he slices the tongue. When I hand him my American Express card, I wait for the comment: *Shapiro your married name?*

At the studio, I add my name to the sign-in sheet. I leave my coat, purse, and deli bag on the bleachers. My stomach growls at the smell of the sandwich, and my mouth waters. The class is packed with dancers, models, fitness-obsessed housewives. The lighting is low, and the moves are jazzy. I like to watch my body, given how hard I have worked to make it skinny. If I squint, I look like one of the dancers. I have remodeled myself, inside and out: the muscles in my arms are defined, and my legs are lean. I was meant to have hips and thighs and breasts, but I have done away with all that. Instead, I'm almost boyish.

Aerobics is about me and the mirror. It's also about sweating. I weigh myself before and after class, and when I finish I'm two pounds lighter than when I began. That these two pounds contain necessary minerals and will cause dehydration is not something I consider. I try to drink as little water as possible. I walk home from the studio light-headed, my ears still thrumming with the disco beat of Donna Summer.

On the way to the meeting, I stop for a manicure. My nails are jagged, the skin around them bitten to the pink. Usually I get something subtle, but this time I go for a color called Gardenia. When I walk out of the Korean salon, my hands look foreign to me. When I studied classical piano, my teacher used to hand me a nail clipper if he could hear a single click against the keyboard. He would be horrified at these perfect, shiny talons. As I walk across Central Park South, heading toward the NBC offices midtown, I have the urge to stop people on the street and tell them that my parents are both in the hospital and neither of them seems to be getting better. I wonder if they can see it on me—the fear, the shame—whether I'm wearing it like a sign on my forehead. All around me, people seem hurried, purposeful. I feel as if a layer of my skin has been peeled back, leaving me naked and raw. As I jostle through crowds of businessmen heading back to their offices after lunch, a thought skitters across my mind: *You're not up for this,* it whispers, but I push it away.

I smile at the guard who gives me a pass to the thirty-seventh floor, take a couple of jagged, deep breaths, and try to let myself fall away as the elevator ascends. There is a cacophony of voices rising in my head, each louder than the next. When the elevator doors open on the thirty-seventh floor, I cannot move. I push my back against the wall of the elevator, grip the rail, and wait until the doors close with a *ding*. I watch the floors light up in descending order. Spots are floating before my eyes. *I can't I can't I can't.*

I walk through the lobby, past the guard, and onto Forty-eighth Street, every cell in my body focused on putting one foot in front of the other. The cold air slaps me in the face, and I breathe it in, hoping to freeze this panic, to stop it from growing. The tension in my body feels as if it's going to shoot through my fingers, out the top of my head. I hear Sheldon telling me to charm the guy's pants off, my mother saying *Eza bat yesh lee*, my father asking me where he is. My own voice is loudest of all, screaming in my head, convincing me that there's no one, nothing that can help me. I can't make it on my own. Like my mother's bones, like my father's brain, I'm going to shatter into a million pieces. There will be no trace of my family left at all.

CHAPTER
FOUR

February. A blustery Sunday. The sky is the color of ash, and the wind whistles and whips across the turnpike. I have begun to drive my father's Subaru back and forth to the city, playing only radio stations he has preprogrammed. I imagine this brings me closer to him, as if listening to Bach cantatas on WNCN, or a sports update, or 1010 WINS News Radio will allow me to enter his interior world—or rather, the world he now inhabits. My father may be a mystery to the doctors, but how can he be a mystery to me? I tell myself that if only I love him enough, I will be able to find the place inside both of us where we share a common language.

And so I try to hear what my father has heard, see what he has seen. I leave the gray velour interior of this car just as I found it. The carefully folded maps inside the glove compartment, the unopened rolls of tokens. I tell myself my father will find everything just as he left it when he gets better. I wonder what he thought about as he drove this car to the train station every morning. I try to recover his thoughts, the images that spun through his mind. Was he happy with my mother? Was he worried about me? Did he see the way I was gulping my wine at dinner, or the way I excused myself from the table and spent long stretches of time in the bathroom? Did Hebrew words and the melody of prayer fly into his head out of nowhere? Did the pills he took make him feel better—or just make him feel nothing at all? The more mangled his brain becomes, the more I try to magically unravel it. The psychic mystery of my father has taken on tangible proportions. Who is the real Paul Shapiro? The gentle, elegant man who used to withdraw behind the gauzy film of self-medication? Or the one who now curses violently and throws his bedclothes across the room?

When I first picked up the car in Tewksbury, an old nubby orange cardigan of his was tossed on the backseat. I have worn it ever since, holding on to it like Linus's blanket in the *Peanuts* comic strip. I am also wearing my mother's wedding band. She has asked me to keep it for her, and so I wear it on the fourth finger of my left hand. I have married my parents for the time being.

A week after the crash, my family has gathered in the lounge on the eighth floor of Overlook Hospital. They are an unlikely assortment of people, together only because they are related by either blood or marriage. Hy, Roz, Morton and Shirley Sugerman are huddled in a corner of the room, gray heads bent together. My father's brother, Harvey, his sister, Shirl, and Susie are camped out near the doorway of the fluorescently lit room, murmuring quietly, breathing into their Styrofoam cups of coffee. Perfect, I think. My mother's family is on one side of this small room, my father's on the other, and I am in the middle.

My cowboy boots have announced me, clickity-clacking down the hall. As usual, I am wearing sunglasses—it's not quite noon, and my eyes are still puffy from whatever I was doing the night before. Memory fails me here. Was I with Lenny at some restaurant, trying to slide oysters Rockefeller down my cocaine-parched throat? Or perhaps Lenny was off getting a chemo-therapy treatment, and I had gone out to a Tex-Mex bar with some girlfriends, where three frozen margaritas landed me cheek-down on the bathroom floor? No matter. I do remember this: I have forgotten to wear gloves on this freezing day—for years, scarves and socks and gloves were for other, more prac-tical people—and my fingers are tinged with blue as I take the warm hands of my family and exchange kisses.

"What's going on?" I ask. I am the only person here who is related to everybody. Basically, it boils down to this: very few people in this room can stand one another.

"We're trying to figure out what to do about Dad," Susie says. "The doctors are saying there's not much more they can do for him here—that he needs a special facility—"

Susie looks like an overgrown flower child this morning—which I suppose she is—bundled in her heavy winter coat, sneakers and leggings poking out the bottom, long blond hair twisted into a messy bun at the nape of her neck.

"What kind of facility?" I am amazed at how quickly my stomach begins to churn, my eyes to swell.

"Rehabilitation," says Harvey, not quite looking at me. I realize with a small shock that his breath smells familiar: an amalgam of vodka and mints.

"Places like Burke in White Plains, or Rusk in the city," says Shirl. She is still holding my hand in both of hers, rubbing it absently. Her pretty face, with its wide flat features and widow's peak, seems to take all this in unblinkingly. One of her brothers is lying down the hall with crossed wires in his brain, her other brother is drunk, and I am standing in front of her wearing no gloves and sopping-wet cowboy boots, reeking of stale nicotine.

Roz walks over to us, wedging herself between Shirl and me. She's wearing a pale yellow sweat suit and improbably bright pink lipstick.

"Dan-Dan," she says. "How's my girl?"

Her face is an inch away from mine, and she scrutinizes me maniacally, just as my mother does. She looks hard and sees nothing. Reflexively, I take a step back. This kind of inspection has always made me wildly uncomfortable. I smile stiffly at Roz, and move to the back of the lounge where Morton, Shirley Sugerman, and Hy are conferring quietly. Like a child, I plop myself onto Hy's lap. He is holding an unlit pipe in his hand, and he smells like tweed and tobacco.

"Sweetie!" he exclaims, jovial as if we were on his La-Z-Boy, settling in to watch a football game. He pats my back, reaches both arms around me and gives me a light squeeze around the middle. Hy is the only one who gets it, I think. I feel him hesitate, about to say something about the way my ribs stick out, my concave belly, but he decides against it. Hy is the one who prescribes diuretics for me. He is an old-fashioned doctor, a surgeon who has been decorated with two Purple Hearts from World War II. Does he have any idea that I take three or four of those diuretics at a time? That I empty out my insides until there's nothing left? It cannot be within the realm of his imagination that I might choose to starve and abuse myself, even as my mother is being fed through tubes and my father fights for his life.

"Let's go see your dad," he says.

Hy and I take the elevator down to the sixth floor, leaving the others in the eighth-floor lounge. My father has been moved to a single room and has private nurses around the clock. The decision to hire outside help was made after one of the hospital staff put him in a straitjacket after he tried to climb out of bed a few nights ago. Ever since, the image of my father thrashing with his arms pinned to his sides has filled my dreams—his face damp with perspiration, eyes wild, his whole body undulating like a giant fish until he finally tires himself out. I have awakened in the middle of the night gasping for air, sweat soaked through my nightgown, hair plastered to my scalp, and for a brief, merciful moment thought this has all just been a nightmare.

My father is slumped in a chair near the window of his room, running an electric razor over his stubbled chin. His glasses are folded on a small table next to him, on top of a stack of magazines, and without them his eyes look close together, angry red marks on either side of his nose.

"Hey, Dad—whatcha doing?" I ask, trying to keep my voice light. I've begun to talk to my father the way I might address a child: slow, cheerful, self-conscious. And he looks at me with the guilelessness of a child, his ability to sense bullshit razor-sharp. Each time I have lied to him (*Mom's not here right now, she'll be back*) he's growled at me to stop it, just stop it! A moment later, he might be laughing, or fast asleep, but in that brief moment of clarity I realize that he knows the truth.

Every time I see him doing something normal—reading, praying, shaving—for an instant my heart lifts and then crashes. I have to constantly remind myself that just because he looks normal doesn't mean he is normal, that if we were to deposit him on the corner of Broad and Wall and ask him to find his way home, as he has ten thousand times over the course of his life, he would be as lost as a puppy.

My father's useless brain, my mother's useless body. Together, they are two halves of a ruined being.

"Hi, Paul-o," says Hy.

"Hello," my father responds vaguely.

"Do you know who I am?"

"Of course I know who you are!"

He looks crestfallen and slightly angry, like a kid caught with his elbow halfway into a cookie jar.

"Okay, then tell me who I am."

"I don't have to." His chin juts out.

"That's fine, that's fine, buddy," murmurs Hy.

Hy pulls a pinpoint flashlight from his jacket pocket and shines it into my father's eyes.

"Paul, do me a favor, just look at me a minute," he says.

My father glares straight at the brother-in-law he has adored for twenty-five years. I watch Hy's face. He moves the flashlight from one of my father's flat green eyes to the other. Hy is the real thing, a talented doctor, and his hunches are often more accurate than the results of all the technology in the world. Behind his soft brown eyes set into their hooded, wrinkled sockets there is something utterly serious, something I cannot name, but which frightens me. (Years later I will wonder if Hy knew it all in that moment: about my father, about himself. Seven months from now, he will be dead of pancreatic cancer.)

I follow him into the hallway.

"What?" I practically beg him. "Please—what? Just tell me." He shakes his head.

"I think Mom and Dad should see each other," he says. He trails off, leaving the end of the sentence unsaid: *before it's too late.*

Shabbos. It came each Friday night, falling over our house at 885 Revere Drive like a pitch-black blanket. There was nothing I could do to stop it. I was not allowed to ride in a car, do homework, play piano, ride my bike, touch lights or any other electrical appliances like toasters, televisions, stereos. My mother spent Friday afternoons cooking meals for the weekend—heavy meat meals like flanken, London broil, pot roast—preparing Friday-night dinner and a cold Saturday lunch, which was my least favorite meal of the week.

Saturdays, after my father returned from shul, my parents and I would sit at one end of the long mahogany dining room table, while on the other side of the window I would see the neighborhood kids flash by on their bikes. I knew they were headed for the woods, where they'd smoke reefer and lie on the damp ground, watching the autumn branches play against the sky. I felt myself twitching under the table, but I knew there was no way I could join them: Instead, I sat quietly in my chair—always the same chair, next to my father at the head of the table, my mother across from me—and dug my nails into the soft flesh of my palm, biding time, my parents' conversation fading away until it was just beyond my reach.

I created a place inside my head, a psychic treehouse. While we ate leftover brisket, sliced challah, and room-temperature vegetables, my parents would talk about Israel—particularly in '72, and again in '76, during the raid on Entebbe—or the stock market, or news of the latest crisis with the Hillside school board. I became good at pretending to listen. It seemed to me that they were really talking in code—about something else entirely. There was tension at the table, a constant abrasiveness between my parents, but when I was around they never, ever, talked about it. I knew some of their battle was being fought over me. My father wanted me to be a good little yeshiva girl, and my mother wanted me to be a pretty, popular teenager. They had agreed when I was born to raise me Orthodox, but that agreement had begun to fray at the edges.

I felt their tension inside me. So I would arrange my features into something resembling interest, and a blankness would settle over me, numb and hazy. It was years before I drank or did drugs, although I knew from watching my father with his pills that there was a chemical solution to the way I was feeling—that someday, when I grew up, I'd be able to find easy ways to disappear. During the years growing up in Hillside, I was there and not there, so far gone that now, when I reach for a concrete memory, some sort of toehold on those Shabbos lunches, my mind fills with white noise just as it did back then.

The Sabbath rules left not too many options: reading was okay, and thinking, praying, walking the dog. We had a black miniature poodle, whose name I'm sorry to say was Poofy, and I would walk him around the block dozens of times, until both of us tired out. Poofy would strain at his leash, and I'd finally take him home, then go up to my room, close the door, and read books I found hidden under my father's night table. The most memorable of these was called *Seven Minutes*. If this book was to be believed—and for a long time I believed—it took exactly seven minutes for a woman to reach orgasm.

After Shabbos lunch, at precisely two in the afternoon, my mother would curl up on the velvet couch in the living room, surrounded by the modern paintings she had collected through the late sixties and seventies: a moody Milton Avery of a green island surrounded by a dark purple sky; a somewhat cheerier Joseph Stella pastel, a strong black-and-white de Kooning in a heavy gold frame. To the left of the couch was a glass case filled with ancient Judaica and pre-Columbian artifacts. She would take off her shoes, rest her feet on the low marble coffee table, and close her eyes. Then, as if by magic, the lights on the stereo would begin to glow, and an afternoon broadcast of *Rigoletto* or *La Bohème* would fill the room. This was made possible by something we called the "Shabbos clock." The Shabbos clock was a switch set on a timer that allowed opera to be played, the television to be watched, lights to be turned on and off at designated hours.

I didn't get it. It seemed to me that the Sabbath was supposed to be a day of rest. It made no sense to me that a timer could turn on the lights but I wasn't allowed to practice the piano. Didn't these rules originate back when illumination involved the hard work of rubbing sticks together? Was playing the piano or riding a bike really considered hard work? I would argue these fine points with my father, but he never argued back. Instead, he hired a tutor to instruct me in Talmud, and to prepare me for my Bat Mitzvah.

The tutor, improbably named Miles, was a rabbinical student at the Jewish Theological Seminary in Manhattan. He took the train out from the city each Sunday, and my father picked him up at the station in Elizabeth. They tootled up the driveway in my father's Citroën, and even from a distance I could see the white crocheted yarmulke perched on top of Miles's dark curly hair. (My father's bald head was almost naked in comparison; he didn't wear yarmulkes except on Shabbos.) My mother greeted Miles with a bit too much noblesse oblige, as if he were a valued but indentured servant. She offered him coffee and a bagel, but he never took her up on it.

Instead, he and I got right down to business. My heart sank as I watched yet another Sunday afternoon bob away, but I knew there was an end in sight. I had struck a deal with my father: as soon as my Bat Mitzvah was over, Miles would be banished. This was something I knew and Miles didn't, which filled me with a secret, evil pleasure. We spread the Book of Ruth over the dining room table, and he sat right next to me, close, so he could point to specific places in the text. He was teaching me to sing the Book of Ruth according to signs that looked like hiero-glyphics, keys to ancient, melodic chanting. Years from now, while walking down Broadway minding my own business, or standing in line at the dry cleaner, I will hear a few notes that will slice through my grown-up self like a knife through the tender meat of my heart.

I hated Miles. His skin was pockmarked, his breath sour, and big flakes of dandruff dusted the collar of his dark jacket. He had no sense of humor—at least none that I could find during these tutoring sessions—and whenever he got excited his left eye twitched. He was everything I feared my future to be, and I was determined to reshape that future at any cost. I didn't know what my options were, but increasingly they seemed to have something to do with my looks. Other people told me I was pretty all the time. I had become aware that the blond, blue-eyed shiksa-goddess fluke of genetics that had Miles and other nice Jewish boys all nervous and flustered was the thing I had going for me. And I certainly didn't think I could rely on my brains: Susie had the brains in the family, after all—and a fat lot of good it was doing her. She was approaching thirty, had a Ph.D., and all my parents talked about was how she wasn't married. They didn't seem to be proud of her academic accomplishments. *She doesn't do a thing to enhance herself,* my mother would complain to my father. *She wears those dreadful glasses.*

The message was loud and clear: I wasn't supposed to turn out like Susie—an unmarried intellectual—and I certainly didn't want to be trapped like my mother by the harsh rules of Ortho-doxy. I was beginning to be interested in sexiness, at least as a concept, and Orthodoxy wasn't sexy. Miles wasn't sexy, with his thick, garlicky breath. My father wasn't sexy when he was dav-ening in shul, but when he put on his suit and got on the train to Wall Street with the other men, something in him changed and became powerful. It seemed fairly obvious: if I was going to have the life I wanted—which was defined entirely by the life I *didn't* want—I would have to focus on being a pretty girl. Somewhere in the upper reaches of Westchester, in a farmhouse surrounded by his wife and children, Lenny Klein was already waiting.

"Smile!"

I am standing next to my mother's bed, aiming a Polaroid camera at her, trying to get her whole body in the frame. She tries to smile, and it makes her face look even more strained than it already is. Her eyes are black pools, cheeks pale despite the blush and lipstick she has applied. These pictures will survive for years to come, and one day, while I am rummaging through her bookcases, a single Polaroid will fall to the floor, and in an instant, time will disappear. I will return to this moment: the harsh winter sun streaming through the tilted venetian slats of her room, the flowers that have begun to pile up everywhere—colorful bunches of peonies, wilted roses, mums, freesia—the clanging inside my head that just won't let up. Since the crash, I have become accustomed to this cacophony. Sometimes I think of it as a voice, other times I imagine it to be pure dissonance, an atonal broken record. I try to drink it away, but drinking only dulls it for a little while, as if it were coming from another room in the same house.

"Beautiful, baby, beautiful!"

I try to joke as I move around my mother, taking her picture from a few different angles. Her nurse, Angie, who is sitting in the corner doing needlepoint, looks up and grins. Angie is in fact beautiful, with auburn curls and a wide, open smile. She is my mother's afternoon nurse. There are three of them—morning, afternoon, and night—and Angie is the best of the lot. The night nurse doesn't do extra things like fluffing my mother's pillows or rubbing cream into the tender places around her wounds. Angie is our angel, the one ray of light in the fluorescent darkness of this place.

The camera spits out green-gray images, blank as television screens, and I flap them in the air until my mother begins to emerge, her black-and-blue cheekbones and thick white casts slowly coming into focus like an image from a dream.

"How do I look?" she asks.

"Like yourself," I say, glancing quickly at Angie, who gives me a thumbs-up sign. "See you in a little bit."

I walk out the door carrying the photos, pausing in the hall to take a deep breath. I have no idea how I'm going to do what I'm about to do. I start moving down the corridor, my feet propelling me toward my father's room. The doctors think that we need to prepare him before he sees my mother, and these photographs are how we're going to do it.

My father is ready. This is the moment he's been waiting for, the only thing he's been able to think about with anything resembling clarity. He's been asking the same question every day— bellowing, begging, weeping, in a white-hot fury—*Where's my wife?*

He is freshly shaven and is wearing the paisley silk robe I brought him from the house in Tewksbury. Although he has lost some weight, he's starting to look healthier. He has some color in his cheeks. He beams at me when I walk in the room, dimples flashing. His nurse—he has not been as lucky as my mother with nurses—barely looks up from her *National Enquirer*.

"Okay, let's go!" he says.

If I could see inside his head, past the swollen blood vessels, the invisible clots forming and dissolving, then forming again, I imagine I would see the sheer power of a twenty-seven-year marriage. He can't remember what day of the week it is, or who's president (when I told him it was Ronald Reagan he looked at me, dumbfounded, and said "No shit!"), but he knows that he has a wife and he needs to see her. This has become the driving fact of his existence.

"Before we go, there's something I need to show you," I say slowly.

He looks at me uncomprehendingly. What could possibly be important enough to cause a delay?

"My wife," he says, his voice rising in a whine. "I want to see my wife!"

He says *my wife*, not *your mother*. I have ceased to exist for him. There is only so much room left in his brain, in his heart.

I hand him the Polaroids, which I have been holding behind my back.

"This is what Mom looks like right now," I say quietly. "She's a little banged up."

He glances at the photographs, eyes sliding over them impatiently. I can tell that he's not taking in what he's seeing. Then, with the grace of a former ballplayer, he tosses them across the room, where they slap against the wall, then scatter.

"Okay," he repeats, his jaw tense and bunched. "Let's *go*."

Hospital regulations have it that he has to be transported in a wheelchair. He submits without a fuss, almost meekly. I arrange a blanket over his lap—as much for his comfort as for fear that his robe will fall open and he'll expose himself—and begin wheeling him down the hall toward the elevator, his nurse walking alongside us, pushing his IV pole.

My father's bald head looks soft, slightly ruined. I bend forward and sniff, as if he were a baby, as if I might inhale some piece of him and never exhale it again. He is focused straight ahead, not looking around at the corridor, the elevator bank, the nurses' station. He is staring into the middle distance. I wonder if he is playing out visions of a sun-dappled day on East Ninth Street, a dark-haired, dark-eyed beauty carrying a hammer on Shabbos.

The elevator doors open and I maneuver his wheelchair inside. Before I have a chance to press the correct floor number, my father has pushed every single button. Then he grins up at me with the glint of mischief I've almost gotten used to by now. I refuse to completely accept that this is my father. I still believe he's going to snap out of it, that any day now I'll walk into his room and be greeted with a warm, grown-up smile and a *Hiya, darling!* and together we will nurse my mother's broken body back to health.

As we pass the nurses' station, I feel their eyes following us—they all know what's going on—but I look straight ahead. I can't stand seeing their sympathy. For years, vital parts of me have been frozen, and since the accident I feel myself thawing, dripping, becoming more human than I can bear. My palms are damp, my heart hammering in my chest, as I push my father's wheelchair through the door of my mother's room.

Her head is turned away from us. She's looking at the corkboard near the window, which is covered with thumbtacked photos I've brought her from Tewksbury: a snapshot of my parents at my mother's graduation from social work school; a picture of my father in a dark suit and tie, a white yarmulke on his head, his face creased into a big sweet smile; the eight-by-ten head shot of me, *Dani York,* hair tousled, eyes unnaturally bright.

"Look who's here!" I announce, trying to keep my voice from quavering.

Slowly, she turns her head. She has applied more lipstick, the coral shade incongruous against her paper-white skin.

"Wahoo," she says weakly, her mouth trembling.

My father is staring at her. His eyes are round. Everything about him is round: eyes, head, stomach, mouth. I push his wheelchair right up to the rail of my mother's bed. I'm tempted to say something, anything, to break this silence, but it feels somehow sacred and essential.

"Wahoo," she repeats like a mantra, a prayer.

My father grabs her hand almost spasmodically, then hunches over and begins to weep, his forehead pressed against the rail.

"How the hell—" he chokes, gulping air.

"Ssshhh," she murmurs. She is crying too, big silent tears rolling from the corners of her eyes, down her temples and onto the pillow. "It's all right now. I'm here."

Does she really think it's all right? My mother has always been a hybrid of sorts, combining equal amounts of terror and optimism. Has optimism won the day?

"We're fucked," my father spits out.

"No, Paul," my mother says firmly. She doesn't flinch, or even raise an eyebrow at his language. And then, with strength I didn't know she had, she releases the rail between them. My father moves closer, and she strokes the top of his head.

"Look at you," he moans. "What happened?"

My father actually seems lucid. Every cell in his body is straining to keep it together, to understand. It has been nearly two weeks since the crash, the longest my parents have ever gone without seeing each other in twenty-seven years. Their hands are entwined, and his head rests gently on her one good shoulder. They cling to each other like two broken dolls.

My mother looks up at me, flushed, triumphant. On her ruined, tearstained face, there is absolute determination.

"You see? We're going to be just fine," she says, her voice fragile as an old wishbone.

On our first date, Lenny Klein took me to the River Cafe, an expensive, elegant restaurant in Brooklyn with sweeping views of the Manhattan skyline. It wasn't until we were halfway there, driving downtown in Lenny's Rolls-Royce, that I allowed myself the thought that I was on a date with Jess's stepfather and we weren't going to be planning a surprise party for her.

At the River Cafe, Lenny handed the maître d' a folded twenty-dollar bill. I had never seen anyone do this before. My father had always made reservations at restaurants and waited patiently at the bar if his table wasn't ready. Lenny and I were led to a window table with a candle flickering next to a small vase of pale pink roses.

"Did you see that?" Lenny asked me once we were seated.

"See what?"

"The way you turned every head in the place. Don't tell me you don't know your own effect."

The truth was that I hadn't noticed anyone looking at me, and I didn't quite believe Lenny.

"Do you know what I told Jess the first time I met you?" Lenny asked huskily, then continued, "I told her you were a golden girl. A perfect angel."

I flushed and looked down at my hands folded in my lap. I didn't know what to say. My last date was with a senior named Adam who bought us a six-pack of Coors and tried to feel me up outside my dorm door.

Lenny produced a pair of bifocals, then skimmed his finger down the wine list, frowning slightly.

A captain appeared at his side.

"Can I be of service with the wine list, sir?"

"Do you have a '59 Margaux?" Lenny asked. "I see the '61, but—"

"I'm sorry, sir, we have only the '61."

"Very well, then."

Lenny leaned back in his seat and smirked at me.

"This wine we're about to drink is older than you are," he said.

I knew what I needed to do. I knew that I would be sunk if I didn't say something—and soon—about Jess, or about his wife. Not to speak up was to become Lenny's accomplice in whatever it was we were doing. But I felt paralyzed, and beneath that paralysis there was a frisson of excitement, an awareness of doing absolutely the wrong thing.

Lenny sniffed the cork and watched the wine being decanted with all the fascination and reverence of watching a ballet. He swirled it around in his glass, took a sip, then gave a nod.

"Pour just a bit for the young lady," he said. "It needs to breathe."

When the captain left, Lenny lifted his glass slightly in my direction.

"To beauty," he said.

I flushed more deeply and looked out the window at a boat moving slowly up the East River. I had an inkling of how much this excited Lenny: young girl, old wine. It was the first time in my life I felt my youth as power. (In years to come, Lenny will turn to me and ask how old I am: twenty, twenty-one, twenty-two, I will answer with a perverse sort of pride, knowing my age acts as an aphrodisiac, knowing that his beautiful wife is too old for him at forty.)

I drank my wine and felt it slide smoothly down my throat, warming the tightness in my chest. I had no experience in the ritual of drinking fine wine. In Hillside, growing up, I thought of red wine as the sweet Manischewitz we sipped out of thimble-sized *bechers* on Shabbos.

After the muscovy duck, after the crème brûlée and cognac, Lenny leaned across the table and ran his finger down my nose in a gesture at once paternal and sexual.

"I'll drive you home," he said with a wink.

Inside his car, I sat all the way against the passenger door as he drove me back to Sarah Lawrence, my cheek hot against the cool window. I felt sick to my stomach. I thought of Jess, back at school. How would I ever tell her I had gone out on a date with her stepfather? Would she ever forgive me?

Lenny fiddled with the dial of the radio until he found a jazz station, a throaty trombone filling the quiet between us. Then, on a long straight stretch of road, he reached down, slowly and deliberately moved my long skirt up my thigh, and squeezed my knee. I knew I should tell him I couldn't ever see him again, but somehow it already seemed too late.

Fast-forward two years. Something has gone wrong, terribly wrong, with my life. I don't, in fact, think of my life as "my life," but rather as a series of random events that have no logical connection. I am no longer a student. I dropped out of Sarah Lawrence after my junior year, supposedly to pursue acting. And I'm actually doing a pretty good imitation of an actress.

But I'm doing an even better imitation of a mistress. Lenny has been busy buying me things. I don't particularly want these things—but they seem to be what Lenny is offering in lieu of himself. So, quite suddenly, overnight, really, I find myself driving a black Mercedes convertible. And just in case I might be mistaken for anything other than a kept woman, I wear a mink coat, a Cartier watch, a Bulgari necklace with an ancient coin at its center. The Mercedes is a step down from the first car Lenny gave me, when we had been going out for a month: a leased Ferrari. I didn't know how to drive a stick shift, so the Ferrari was a bit of a problem. What I must have looked like! A twenty-year-old blonde dressed like Ivana Trump, stalled out in traffic, grinding gears, trying to find the point on the clutch to hold that ridiculous car in place.

My parents know something is up. I am drifting away from them—and they are letting me. They know I'm going out with *somebody*. Lenny rents an apartment on a pretty little street in Greenwich Village, a furnished triplex with a garden, a fireplace, and a bedroom with a four-poster bed. He calls it "our house," as if he doesn't have another house with a whole family in it an hour north of here; he keeps a half-dozen suits in the bedroom closet and a brand-new silk robe hangs behind the bathroom door. There is an entire floor we don't use, consisting of a large, airy children's nursery.

One night, I invite my parents over for dinner, and I remove all traces of Lenny from sight. Of course, there are clues: a glossy brochure for Italian yachts; a humidor in the center of the coffee table; a man's Burberry overcoat hung on a hook near the front door.

I have cooked up a storm, and the place is filled with homey smells. Garlic, basil, coriander. It is winter, and snow is piled up on the sills. Spotlights in the backyard shine on the landscaped garden, the redwood table and Adirondack-style chairs, and huge terra-cotta pots of last spring's dead geraniums. I have my father's favorite music—Dvořák's symphony *From the New World*—playing on the stereo system.

My parents ring the doorbell. They look so solid standing on my front stoop, their cold, red noses poking out from above their mufflers. If nothing else, my parents look as if they belong together. They are elegant and rangy, similarly proportioned. (Unlike Lenny and me. Lenny is thick as a linebacker, and I have become so delicate the wind may pick me up and blow me away.) My mother strides into the brownstone as if this weren't the weirdest thing in the world, visiting her daughter in a lavish apartment with no name on the outside buzzer. My father trails behind her warily, as if he's setting foot on another planet.

My mother sings. She enters the living room, taking up space, flinging her arms wide and doing an impromptu little dance to the Dvořák symphony.

"Tra-la-la-la," she trills.

I cringe inwardly, grateful that it's just the three of us, that no one is here to witness this display. My father and I hang back and watch, our faces crumpled into awkward little smiles. We're used to it, after all. In every family, there is room for only one Sarah Bernhardt, and in ours, my mother has assumed that role. I don't realize that my mother is frightened—that this is a lot for her to take in, her college-dropout daughter living in the lap of luxury. All I see is her outsized self, twirling around my living room in her fur coat and boots.

All of a sudden, I want a drink. I walk over to my mother and put a hand on her shoulder, and she spins to a halt. I take her coat, and my father's, and hang them above Lenny's raincoat by the front door. For the first time, I notice that there's a wreath made of twigs, a bit of Americana, on the wall near the kitchen, and I wonder if I can remove it quickly before my father sees it. Wreaths, under any circumstances, are as goyish as it gets. Which would be worse for my father? Imagining I'm with some powerful guy old enough to be my father? Or the possibility that the guy isn't Jewish? *Yes, Daddy. He's Jewish. Twenty-three years older than me, a pathological liar, married to a woman who knows nothing of me—and a Jew.*

I pour my parents two glasses of Chardonnay and a large vodka for myself. I figure if the vodka is in a water glass they won't know the difference—especially if I drink it as if it's water.

My drinking has taken on a new urgency in the past few months. It's no longer a question of desire but of need. I cannot get through an evening like this without the armor of booze. I hand them their wine and direct them to the couch. On the coffee table, I have put out a vegetable plate and a bowl of olives.

"Quite a place," my mother says brightly, her gaze darting around the room at the white brick fireplace with its wrought-iron tools, the glass wall overlooking the garden, the soaring ceiling. My father stares at the fringe of the rug, glassy-eyed. He needs to be as numbed out as I am to get through this night.

"Thanks," I murmur, as if she's paying me a compliment.

I check on dinner, using the opportunity to gulp some wine from the open bottle in the fridge. Vodka and white wine is a combination I know works for me. If I just stick with the formula, things shouldn't be too bad in the morning. It'll only become a problem if I switch to red or have cognac after dinner. I've learned to color-code my booze: clear (vodka, white wine) and colored (scotch, cognac, red wine) shouldn't be mixed. Especially if I'm not eating.

I've prepared my signature dish, which has become my signature because it's literally the only thing I know how to cook. A recipe out of *The Silver Palate Cookbook*, it's a chicken stew of sorts, with white wine, olives, prunes, and brown sugar. I'm serving it with wild rice and a string bean casserole I bought ready-made at Balducci's. For dessert, a tarte Tatin from Patisserie Lanciani. I have run all over the West Village preparing for this evening, thinking that maybe my parents will be impressed by my culinary efforts, so impressed that, by the end of dinner, patting their full stomachs, they'll swell with pride at their only daughter who is, after all, living such a gracious and well-appointed life.

"Can I help?"

My mother is standing in the doorway. How long has she been there? Did she see me take the swig from the bottle of wine? My mind races with how to explain it. *Thirsty* is the only word I can think of. But then I realize that she hasn't seen anything at all.

"Actually, I think everything's under control," I say, carrying the casserole to the table, which I have set with linen place mats and napkins. In the center of the table, there is a vase of drooping purple tulips.

The silverware, the pots and pans, the linens are all courtesy of the owner of this sublet place, a woman whom I might view— if I were thinking of such things—as a cautionary tale. A blond, stylish, whippet-thin, fiftyish real estate broker, she has lines around her mouth that aren't from smiling. She occasionally stops by to fix something in the garden or the basement, and when I get near her I smell vodka and stale nicotine just beneath a cloud of L'Air du Temps.

The music has stopped by the time my parents and I sit at the dining room table, but I don't notice. If I did, I would certainly change the tape, fill the air with something other than the tinny, lonely sound of our three forks scraping against plates. I push my chicken from one side of my plate to the other, my stomach clenching and growling in protest. I have allowed myself a glass of wine in front of my parents, using one of the crystal wineglasses Lenny bought me as a housewarming gift. It's all I can do not to gulp it.

It seems my parents and I, after twenty-two years in one another's company, have run out of things to say. They're not talking about the political situation in Israel, and we can't discuss my schoolwork, since I'm not in school. My father presses a corner of his napkin to his lips, and murmurs something about the food's being delicious, and my mother energetically concurs.

"My wonderful daughter," she says, shaking her head. "You've turned into such a little homemaker."

I look at my parents across the table. Is that what they really think? How can they just sit there? Some small piece of me wants my father to fling me over his shoulder and carry me, kicking and screaming, to their car parked outside. I secretly wish they would drive me home to Hillside, deposit me in my childhood bedroom and feed me chicken soup and Saltines. I want to start my life over again, but I don't know how.

I'm afraid I'm going to cry, so I walk into the kitchen and pull the apple tart from its box, arranging it on a cake platter. What did I expect from this evening? I thought I wanted my parents to be proud of me, to see that I'm living like an adult. But even I know that isn't true. We're all playing a game here, pretending this is a nice family moment: mother, father, and daughter eating an elegant meal.

I present dessert with a flourish. The tarte Tatin, and espresso fresh from the brand-new espresso maker. Finally, the conversation—or lack thereof—veers, like the tide, in the only direction it can.

"Can't you tell us who he is?" my mother asks as I take one bite of the delicious, flaky apple tart, then another, and another. I'm ravenous like a starving dog. I'll make myself throw up later.

My father clears his throat. "It's been so long, Dani—it seems we really ought to know—"

I keep shoveling pieces of apple and crust into my mouth. I can actually feel my stomach closing around each morsel of food. My parents have been strangely passive on the subject of my dropping out of college and taking up with a mystery man. Since they are no longer supporting me, perhaps they feel they have lost parental control. To me, it just feels as if they've given up on me. Over the past year I have returned from trips to

Europe bearing gifts for my parents: Charvet silk ties for my father, brightly printed Pucci scarves for my mother. They are gifts I couldn't possibly afford to buy on my meager income from television commercials. Who do they think paid for those gifts? And why do they accept them? Part of me is screaming to tell them already, just get it over with. After all, they've met Lenny on Parents' Day. They know who he is. Has he crossed their minds as a possibility?

His name is on the tip of my tongue. It would be so easy. *It's Lenny Klein,* I could say, then watch the chips fall. Would they be horrified—or relieved? What can they possibly be imagining? I am woozy from the vodka, wine, and two helpings of apple tart. *Okay,* I think to myself. *Okay, just fucking say it.*

"Is it Teddy Kennedy?" my mother asks. I can tell she's really considered this. She looks at me eagerly. Does she want it to be? An image of the bulbous-nosed, red-faced senator from Massachusetts flashes through my head. My mother is staring at me, wide-eyed, poised for an answer, and I can't say anything at all.

It is two and a half weeks since the accident, and Lenny has talked me into going with him to London. The Concorde can have me back in four hours, he reasons, and it seems my parents have both stabilized. I don't have to be here every minute of every day.

"Look at you," he says. "You need a break."

He points out the dark circles under my eyes, my pale, drawn face. On my upper arm, where I was given an injection of cortisone for the allergic reaction, my flesh has withered into a fist-shaped indentation, as if someone punched me hard and I just haven't bounced back. He says I may even have become too skinny—something, for Lenny, almost unimaginable. I prefer to see myself as newly tough, a featherweight fighter blasting her way through this terrible winter on sheer grit.

I tell myself that a break is just what the doctor ordered. That taking care of my parents every day for two and a half weeks has been a valiant effort, and that I need—deserve!—a vacation. Roz can spell me, I think. Or Susie. I have become resentful of Susie in the last weeks, the way the sick resent the healthy. She dashes into the hospital at the end of her workday, her skin dewy, flushed from the drive out, important-looking papers stuffed into a heavy briefcase. These papers must be notes on her patients, case histories, life stories condensed into a few terse diagnostic sentences. I can only imagine what my half sister would jot down about me: *spoiled brat,* most likely. *Wish she'd never been born.* I think it's safe to say Susie and I are not getting along. We are each playing our roles perfectly: she's the responsible, serious older sister, and I'm the flighty, unreliable younger one.

Except that the lion's share of taking care of my parents has fallen on my shoulders. My mother is not Susie's mother, after all, and our father seems to have lost all sense of time. Three days can seem like three hours to him. He doesn't keep track of our comings and goings. All he seems to be able to focus on is when he's seeing my mother. My parents have established a routine over the past week. At lunchtime my father is wheeled into my mother's room, and they eat together, their trays side by side. My parents used to love fine restaurants, particularly fish restaurants where they were able to eat kosher meals with little compromise. Gloucester House, Le Cygne, Sea Fare of the Aegean. Now they sit in their green cotton pajamas, my mother removing the cellophane from the top of my father's Jell-O.

And so I actually manage to convince myself that a trip to London is a perfectly reasonable thing to do. It will not be the first time I've flown on the Concorde. Lenny and I have zigzagged across the Atlantic a half-dozen times; for years to come I will find British Airways and Air France shaving kits, slippers, toothpaste buried in back drawers like loose glitter.

The morning we are supposed to leave, I drive out to Overlook to say good-bye to my parents. It is a Sunday, and the sky is a bright, almost iridescent blue—a blue I will remember. I will also remember what I am wearing on this particular morning. To fly to London on this February day, the twenty-third of February 1986, I am wearing a silk blouse borrowed from my friend Diane, a Sarah Lawrence girl who works in a trendy Soho boutique where she gets discounts on high fashion. The blouse is notable for its color (gold) and design (slit all the way up the back). The blouse is tucked into a short black skirt. I don't remember my shoes, stockings, or bag. I assume there's a winter coat in the picture, but this may be assuming too much. I can supply the moment with bright lipstick, to be sure. Red, perfectly outlined lips painted on the blank palette of my face.

When I arrive at the hospital, I go through the same choice I've had to make each day since the accident: Do I go to my father first? Or my mother? I have solved this problem—or at least alleviated my guilt about it—by alternating between them. Today is my father's turn, but something makes me press the elevator button for my mother's floor. I figure I'll visit with my mother, then go to my father. It will be easier to say good-bye to him, since he won't really notice that I'm leaving.

The door to my mother's room is closed. There is a glass pane in the door, just like the one we had in Hillside—only instead of seeing a marble foyer with an old Spanish table covered with piles of magazines and mail, instead of hearing the click and yap of Poofy, I see my mother in her hospital bed, surrounded by strangers.

I push the door open.

"There's the daughter," someone says. Or am I imagining this? Has this memory been thickened by successive years of memory, each like a coat of paint obscuring the colors beneath?

There's the daughter.

Somebody get a chair.

Where's the other daughter?

I open my mouth but no sound comes out. In the instant before my mother reaches her trembling arms out to me, I already know. These people are clergy, social workers, nurses. They are here because something terrible is happening elsewhere, on another floor of this hospital.

"Come here, Dani," my mother cries. I go to her, climbing between her casts, trying to get as close as I can without hurting her. I hold her head against my chest.

"It's okay," I whisper, stroking her hair. Her whole body is shaking. I remember one of the doctors telling me that my mother isn't out of the woods yet; with the number of fractures her body has sustained, any kind of shock could literally kill her.

"Ssshhh," I breathe into her ear. I wrap my arms around her, holding her like a newborn. Our hearts pound together.

I look up at a woman standing by the foot of my mother's bed, a stranger in a multicolored woven vest and ethnic jewelry.

"What happened?" I ask.

"Your father—they're working on him," she says.

"What does that mean?" My own voice sounds as if it's approaching me from far away, shrill as a siren.

"They're doing everything they can."

As I kiss the top of my mother's head I am aware that if it were a couple of hours later, I'd be halfway to London. I'd be drinking scotch on the Concorde, watching the digital display in the front of the cabin where you can see the precise moment you break the sound barrier.

All I know right now is this: I pressed the elevator button for the sixth floor rather than the eighth, and for that reason alone I am not standing in the doorway of my father's room watching electrodes be placed against his chest, watching them cut open his throat and place a tube inside, hearing shouts and the pounding of rubber-soled feet. I am not watching my father die.

CHAPTER
FIVE

The morning of my father's death seems to be a moment I have been preparing for all my life. I sink into a cold, stark numbness as if taking a single step backward into my own shadow. As we sit and wait—my mother, the social worker, the hospital priest, two nurses, and myself—it seems inevitable. *Beshert*—a Yiddish word that loosely translates into *meant to be*—ricochets through my head. I gird myself, preparing for the final blow, as if the shape of grief is something I am familiar with and I know it will knock the breath out of me.

It is just past ten in the morning when the doctor walks through the door of my mother's room, his head bowed. *Sorry*, he says, then *embolism*, and then, *autopsy*. The next few seconds slow to a standstill as I cradle by mother in my arms, as if I can shield her from this information.

"I'm here," I murmur into her ear. "I'm here," I repeat, as if my presence can possibly be of comfort. She is trembling all over like a wounded animal, and I am a crevice into which she has crawled to seek shelter.

I'm saying this as much for myself as for my mother. In years to come, I will be grateful for the small kernels of luck embedded into the center of this horror show: I am not in London with Lenny. They will not have to postpone my father's funeral in order for me to return home.

"Doctor, there will be no autopsy. It's against our religion." My mother's voice is amazingly strong, but there is a crack in the middle of it, a fissure. She isn't even Orthodox. How does she know that the body is considered sacred, that it is a sin to cut into it? Her face is absolutely still, collapsed into itself.

I call Susie.

"What's up?" she asks breezily. For the second time in just a few weeks I am in the position of being the bearer of shattering news. I imagine my half sister in her West Village apartment, the morning sunlight gleaming against her grand piano, the Sunday Arts & Leisure section spread across her coffee table.

Do I tell her to sit down? Is it possible to fall back on such a cliché? Susie would remember the moment better than I do. Time has supplied me with a gauzy scrim, except for this: as I tell my half sister, just saying it straight—*It's Dad, Susie*—I will not let go of my mother's hand. I am afraid she is about to slip away from me. The possibility of her death has its own presence in this room. The doctor has moved to the other side of her bed, where he is holding her hand by the wrist, surreptitiously taking her pulse.

Susie's voice barely changes. Years of psychoanalytic training serve, in a moment like this, as a way of keeping her reactions—if not emotions—in check. Whatever actually goes on inside of her, she doesn't miss a beat.

"Where are you?" she asks.

"I'm here—with Irene."

"Do Harvey and Shirl know?"

"No. I called you first."

"I'll call them," she says. "Shirl and Moe will need to get down from Boston. I'll call Harvey at Gram's and drive out to the hospital with him."

"Okay," I say slowly. I don't want to let her off the phone. I want something to happen between Susie and me—a moment of shared grief—but instead I feel the distance between us, stretched as tightly as the phone's wire that I have wrapped around my wrist.

"Dan, make sure they don't touch Dad," she says.

"What do you mean?"

"No autopsy—"

"Irene already took care of that—"

There is an ironic little pause in which, without saying a word, Susie and I both know that we are contemplating my mother's less than encyclopedic knowledge of Orthodox ritual.

"Someone has to stay with the body," she says.

It should not surprise me that Susie knows more about this than I do.

"Okay," I say faintly. "I'll take care of it."

I am tracing the ropey veins of my mother's left hand as I catch the doctor's eye. He inclines his head at me, then at the door, letting me know he wants to talk to me in private.

"Mom, I'm going to get something to drink," I whisper.

"Don't leave—" she moans.

"I'll be right back—I promise."

I follow the doctor into the corridor. The hospital has suddenly become as strange to me as the set of a science fiction movie. Where do bodies go? The thought that my father is being wheeled into some hospital morgue is more than I can bear.

"Do you want to see your father?" the doctor asks.

This stops me in my tracks.

"You mean—" I falter.

"Before he's taken—"

"No," I say, squeezing my eyes shut. This is a moment I can never take back, and yet I am certain that I don't want to see my father dead. If I see him that way, I imagine that image will become all I will ever be able to remember.

"Can you just—ask them to leave my father in his room until his family gets here?" I ask. "It's against our religion to move him."

I have no idea if this is true, but it certainly seems effective. The doctor stops at the nurses' station and confers with one of the nurses who nods her head and picks up the phone. Then he turns to me and I see that he feels bad about this, he really does. He knows that I am looking at him thinking: This is the man who could not save my father's life.

"Miss Shapiro, your mother—"

He hesitates.

"Go on," I say. "Whatever you have to say, just say it."

"Your mother is a very sick woman. I know she's been looking better, but she's nowhere near out of the woods."

He pauses again, looking at me searchingly. What is he trying to tell me?

"She may not survive this," he says. "We have to keep her as calm as possible."

I have always believed my mother would live forever. While I have never said good-bye to my father without the thought crossing my mind that I might never see him again, my mother has seemed indestructible, fixed in my consciousness like a gnarled and stately tree that has taken root there. If she is ripped away at this moment in my life, she will take her roots with her and I will be left with less than nothing: a ragged, empty hole.

"My mother will survive," I say to the doctor forcefully, as if the depths of my need for this to be true will make the slightest difference.

It seems no time at all has elapsed before Harvey and Susie have materialized in the corridor. At least an hour has gone by—it would have taken that long for Susie to go to her garage, pick up Harvey, and drive from the West Village to Summit, even with no traffic—but it seems time has stopped.

Susie doesn't look me in the eye.

"We were upstairs to see Dad," she says accusingly. "They intubated him."

I don't know what this means. Later, I will find out that they opened my father's trachea and tried to get him breathing again.

"Harvey had to actually pull the tubes out himself," Susie says. She is pale and looks exhausted, and is wearing layers of old cotton sweat clothes that look slept in. She must have grabbed her keys and walked straight out of her apartment when she got the news. I want to hug her, to reach out and hold her. She is all I have left of my father, and I want to breathe her in and see if anything of him remains.

"Have you been to see Irene yet?" I ask.

"No."

Harvey, who looks like a slightly Satanic version of my father—olive-skinned, with a pointy dark beard—seems kind of crazed. He is pacing back and forth, running a hand through his thinning hair. His eyes, through the thick lenses of his glasses, are red and swollen. He has always been envious of his more successful, more respectable older brother. What was it like for him to pass his hand over the eyes of that brother, closing them forever? Has he been drinking? Harvey's alcoholism, as I will later learn to call it, is not something I think about on this particular morning, but I have no doubt it impacts every step he takes.

"I called Riverside," Harvey says, referring to the Upper West Side memorial chapel. "They scheduled the funeral for eleven tomorrow."

"The Chevre Kadishe are coming to take Dad," Susie says. "There are some papers we need to sign."

"What's the Chevre Kadishe?" I ask. Another couple of words I've never heard before. My vocabulary is expanding by the minute.

"They make sure the body is prepared properly for burial," Susie says.

"Has anyone gotten in touch with Rabbi Riskin?" Harvey asks.

They are talking over one another, at me, into the vast, swirling whiteness of the corridor. We are twenty paces from my mother's room, and my uncle and half sister are making decisions that it suddenly dawns on me are not theirs to make.

"Why don't we go talk to Irene about all this?" I ask, my voice thin with rage. They both look at me, startled. It seems my mother has not been a factor in considering arrangements for my father's funeral.

We march into my mother's room. She has her little family around her—Roz and Hy, Morton and Shirley—and I am relieved to see her taking small sips from a paper cup of orange juice. She is clutching a photograph of my father, the one that had been tacked to her bulletin board. I see his wide, dimpled grin, his eyes cast to the side, the white yarmulke covering the dome of his head.

"Hello, dear."

"Oh, Harv—" my mother clasps her arms around Harvey's neck and draws him close. Her eyes are wide and soft. On this day, my mother and her in-laws have something in common that transcends the bitterness and misunderstandings between them.

Susie bends down to give her a kiss.

"Irene, I'm so sorry," she murmurs, her hair falling over both their faces. The life that Susie and my mother shared, which began when Susie was nine years old, has now abruptly ended, though it will be months before it becomes clear that they will never speak again.

"Oh God, Susie—"

At the sight of my half sister, my mother begins to cry. She seems surprised to see each new face looming over her bed. Out of the corner of my eye, I see Roz pat Harvey's back, her hand moving in figure eights. Morton and Shirley Sugerman are standing in the doorway, and Hy is sitting on the edge of my mother's bed. He pulls a hankie from his jacket pocket and presses it into her clenched fist. The room is crowded with people who never would have known one another if not for a chance meeting between a man and a woman on East Ninth Street thirty years ago.

"Irene, if it's all right with you—" Harvey begins, his eyes flickering toward me, "I called Riverside to make arrangements."

"Riverside," my mother repeats, her brow knitting together. "But that won't be possible."

"Why not?" Susie asks.

"Look at me," my mother says, gesturing to her legs, casts dangling in traction. A few signatures are dotted across the plaster, along with scattered little messages. Someone actually wrote *Get well soon*.

"I can't be moved."

"So . . . what are you suggesting?" Harvey asks slowly.

"The funeral's going to have to be here in the hospital," says my mother.

"But—" Harvey begins.

Roz removes her hand from his back.

"If that's the way Irene wants it, that's the way it's going to be," she says sharply.

"The hospital isn't exactly set up for a funeral," says Susie, using the neutral voice she probably uses to talk down psychotic patients. "Don't you think we ought to—"

"I'm going to attend my husband's funeral," my mother says. Her chin lifts, her eyes darken, and in her face I see absolute determination. If she had to get out of bed and walk there, at this moment I believe she would and could.

For the first time since I have known him, I call Lenny's house. Or to be more precise, I call his house and don't hang up when someone answers the phone. For years now, I have grown accustomed to the sound of his wife's hello. She's one of those people who divide the word into three happy syllables: *hel-lo-o*, she practically chirps into the phone. Usually I just hang on and breathe quietly, cradling the receiver against my shoulder. I listen as her hello grows puzzled, then impatient. *Is anybody there? Who is this?*

This time a child answers the phone. Lenny has an astounding number of children: three stepdaughters (including Jess), three adopted daughters from an earlier marriage, and two daughters with his wife. They range from my age to the toddler.

"Hi, is your daddy there?" I ask.

"No." She says it softly, almost inquisitively.

"Okay, thanks."

Shit. Lenny's probably on his way to the airport. Suddenly I remember that one of his partners is traveling to London with us and think maybe I can catch him at home before Lenny picks him up. I dial directory assistance for Mount Kisco and get the partner's home number.

His wife answers the phone.

"Hi, this is a friend of Lenny Klein's," I say. "I'm trying to reach him. Have they left yet?"

"No, they're right here," she says. She sounds puzzled, and why wouldn't she?

Lenny gets on the phone, his voice furious, but careful.

"Is there a problem?" he asks. We both know I have just broken the cardinal rule.

"Sorry to bother you, but my father's dead," I say evenly.

Silence. I can almost hear his mind clicking.

"Oh, Fox," he sighs. "I'm sorry. What do you want me to do?"

What does he *mean*? If I were his wife, would he be asking what I want him to do? I feel like telling him to just go to London. After all, what place does he have in what comes next? My extended family has never met him, my half sister can't stand him, and my mother will, to say the least, not find his presence comforting. But I feel that I need him. My mother is alone, Susie is alone—but I cannot bear to be alone.

"Get here as soon as you can," I say.

He pauses. Was this what he was expecting?

"I'll be there within the hour."

In the meantime my mother has become a whirling dervish of organization. From her hospital bed, legs dangling above her, she orchestrates the details of her husband's funeral. My mother is a formidable woman, and no less formidable for being flat on her back. No is not an option, she tells the hospital administrators who balk at the idea of two hundred people descending on their chapel tomorrow morning. She calls Isaac Swift, the great Orthodox rabbi who married my parents almost thirty years ago, and asks him to perform the service. She hands me lists of names and numbers, some unfamiliar to me, and asks me to begin breaking the news. From now on, when I think of the morning of my father's death, I will picture myself standing by a pay phone in the Overlook Hospital corridor, wearing a gold silk shirt slit up the back, a black miniskirt and heels, crossing names off a yellow legal pad, the shocked moans of my father's old friends echoing in my ears.

Lenny arrives at Overlook at the same time as my friend Diane. They walk quickly down the hall toward me, their noses red from the cold, bundled into scarves and overcoats. Diane, whose mother died when we were freshmen at Sarah Lawrence, is my only friend who understands what it's like to suddenly have to grow up. She reaches me before Lenny does, and gives me a long, close hug. I close my eyes and try to breathe. Diane's wearing a floral perfume, her hair is soft against my face, and suddenly I am weeping. It has only been a couple of years since we were college students together, and never have I so desperately wished I could turn back the clock. It seems as if everything that's happened since I left college has slowly, inexorably led to this moment. I imagine that all this is somehow my fault. I caused my parents' accident as surely as if I had been behind the wheel of the car on that snowy night two weeks ago.

"Let it out," she whispers.

Huge sobs rack my body, and I feel Lenny's hand tentatively pat my back. I hate him, and I hate myself. Some small part of me knows that I will always be horrified by Lenny's presence here. That he will be forever wrapped up in my memory of this day, like a blurry face in a grainy old group photo, under which a caption reads: *Unidentified man, second from left*.

As Lenny, Diane, and I leave the hospital to get some lunch, my father's sister, Shirl, and her husband, Moe, pull up in a taxi from the airport. Shirl's face is etched with grief, the lines around her forehead and under her eyes more pronounced than I've ever seen them. Her eyes are filmy slits. Moe, who looks something like Leonard Bernstein in a yarmulke, is holding her by the elbow, and they both seem frail and old. The whole world has tilted off its axis overnight. Violence is in the air—violence and randomness. If my father had passed out not behind the wheel of his car during a snow storm but at lunch at the stock exchange, or while pulling on his socks in the morning, things would now be different. He might still be alive. And he would not have very nearly taken my mother with him.

Shirl gives me a hug, and I introduce her to Lenny. It's a little like bringing Queen Elizabeth around to visit a minimum-security prison. These are two people with nothing in common. Still, she grasps his hand with both of hers and gazes at him with deep sadness, as if his presence here must mean that he's virtually a member of the family.

Lenny, Diane, and I head into Summit for lunch. The town's streets seem to be filled with families: young parents pushing toddlers in strollers, their breath vaporous clouds in the cold, shepherding gray-haired, pink-faced grandparents into the doors of restaurants. Christmas decorations are still on the lampposts and trees, and there are sale signs in all the store windows. As we wait for a table, a collective cheer goes up as a goal is scored in the hockey game being broadcast on the television at the end of the bar. My father is dead, and the Rangers are winning.

I don't know whether I can stomach any food, and though I tell myself I'm going to order a Diet Coke, I find myself asking for a screwdriver. The drink comes in a tall glass, and I can tell from the color of the orange juice that there's a healthy shot of vodka in there. Could the waitress tell how much I needed it? Did she tell the bartender to make it a double?

I gulp the screwdriver, then order another, this time asking for a double. No one suggests that this might not be a good idea; no one tells me to stop. Lenny holds my hand, stroking my fingers, fiddling with the diamond friendship ring he gave me last Valentine's Day. I feel as if my eyes have been stripped. I horrify myself with my next thought: I wish he were dead.

I excuse myself, saying I'm going to the ladies' room, and instead stop at the bar and knock back a quick shot. When I am fortified enough, numbed against my own numbness, we head back to the hospital.

Many years from now, when my father is a skeleton in the ground, when my mother strides the streets of New York City with her arms swinging, when Lenny Klein is reduced to a colorful if painful story in my mind, there will occasionally be a day when I feel the fear. The knot inside me will unravel, and suddenly my heart will pound uncontrollably, my palms will dampen, and my ears will begin to ring. Wherever I am, I will be desperate to escape. The urge will be to run as far as I can, as fast as I can, away from my own body. Like a tidal wave, it will come out of nowhere, this nameless, faceless terror.

I will try every so-called panacea, every cure—meditation, homeopathy, behavioral therapy—but nothing will make it go away. Not *really*. Sometimes, I will be lulled into thinking it's gone for good. But then years will pass, and suddenly it will be there again, haunting me like an old lover. It will return to remind me that I am my father's daughter. That I inherited his terror, along with my mother's will to survive.

It is the winter of 1990, four years after my father's death. I am at the Metropolitan Opera, clutching my mother's elegant little opera glasses. My mother has lent me her subscription seat for the evening, and it is perfect: right center aisle, eight rows from the curtain. The lights dim, the curtains rustle, and the dramatic opening notes to *Tosca* fill the darkness.

And then it begins. What sets it off? Perhaps it is the older couple in front of me, their gray heads bent together over the program. They remind me of my parents and how they are not growing old together. Or maybe it's my own sense of myself in this moment, as a young woman at the opera who looks as if she doesn't have a care in the world. On the outside, I am cool and impassive, but inside me, the resounding chords have become indistinguishable from the pounding of my heart, which is suddenly racing as if I've just done the hundred-meter dash. The curtain is rising, but I can barely see a thing. My vision has gone blurry, and my fingertips and toes are tingling. A thousand thoughts race through my mind, a collage of images. Is there a doctor in the house? Surely, in this opera-going crowd, there are at least a few cardiologists. In quick order, I imagine the production grinding to a halt, my body slumped in the plush velvet seat. I am still in my twenties, and I am picturing my own death.

Quietly, I reach two fingers beneath the silk scarf encircling my throat, and take my own pulse. The image of my father—the father of my childhood—flashes before me. He is hunched over at the kitchen table holding his wrist, checking his pulse, his mouth tight. He taps a few pills into the palm of his hand and downs them in a single gulp. I follow the second hand on my watch for fifteen seconds, and realize my heart is pounding at nearly one hundred and forty beats a minute.

I'm dying, I think to myself. *Either that or I'm going crazy.* I close my eyes against the swirl, waiting for the wave to crash over me, to sweep me out to sea. I have been expecting this; it does not surprise me. I am desperate for relief in any form—a handful of pills, a shot glass of scotch. This must have been what my father felt all his life, and he has passed it on to me in a sharp and terrifying legacy. There is a phrase running through my head, but the din is so loud it takes me a while to make it out.

Of course. It is my father's voice. *Of course.*

The morning of my father's funeral, I take a car service from the city to Overlook. The driver is the same guy who picked me up at the airport when I got back from California, but this time he doesn't try to make conversation. Lenny is with me, and my friend Annette, whom I met in acting class. Annette hates Lenny. She's a pretty girl from Texas with big round eyes and a flat drawl, and has a no-nonsense attitude about flashy married men who fuck around—but she's also afraid of him. To make ends meet, she does secretarial temp work and has recently been employed by Lenny's law firm. He doesn't know this, and she has sworn me to secrecy. One drunken night in the near future, I will break my promise to Annette and tell Lenny that she temps for Belzer, Klein, Marchese & Rosenzweig, and he will march down to Personnel and see to it that she is fired. But today I am flanked by them in the backseat of the town car, holding a paper pint of Tropicana, trying to sip orange juice through a straw. I am afraid I'm about to black out.

"We'll get you a real drink as soon as we get there, Fox," Lenny says.

"She doesn't need that—" Annette blurts out.

"If you don't mind, I think I just might know what she needs better than you do," says Lenny, staring her down.

Annette flushes.

"But she really ought to keep her head clear—"

"Will you both just stop it!" My voice is hoarse from the two packs of cigarettes I smoked the night before, and I can't breathe in all the way. Yesterday's lunchtime vodkas and the bottle of wine I drank for dinner have left my hands shaking. I focus on the back of the driver's neck, pink above his white collar. There are two photographs Scotch-taped to the dashboard—I don't remember noticing them before—class photos of a little boy and girl smiling gummily against a sky-blue backdrop. They look happy. I stare into the girl's eyes and try to see her future: will she someday be riding in a limo next to her married lover, on her way to her own father's funeral?

I lean my head against Lenny's shoulder. He smells of after-shave and starch. He cannot help me. No one can. I feel myself begin to float away, and pull myself back with an invisible thread, a single, delicate strand of sanity. My entire life has been distilled into a single pinpoint of purpose. My father is dead, and I stopped caring about myself long ago, but today I have a mission: if I let go of the thread, I know that my mother won't have a prayer in hell of surviving.

She is propped up in bed, wearing a dark burgundy silk robe over her hospital pajamas. I will later learn that my friend Diane brought her the robe from the city, that it belonged to Diane's boyfriend's grandfather. These are the kinds of details that remain fixed in my memory—their very oddity is what makes them indelible. For my father's funeral, my mother is wearing the clothing of a stranger.

"My beautiful daughter, all dressed in black," she murmurs, reaching her arms out to me as I walk through the door to her room, Lenny and Annette trailing behind me. My mother's propensity to see things in pictures—a trait I have inherited—has not left her, even in this moment. There are people, all sorts of people, milling about. Doctors, relatives, friends. Someone has run a comb through her hair, and she is wearing lipstick.

"It's time—" someone whispers.

A stretcher is wheeled next to my mother's bed. It takes three orderlies to lift her, and as they do I see the tension in their faces, as if she were a piece of delicate china only just glued back together. As if she can shatter into a thousand pieces on the floor. A nurse carefully maneuvers the intravenous pole and traction equipment around as they place her on the stretcher.

Once she is settled, I realize I have been holding my breath. I try to swallow, but my throat is dry. I take her left hand and gently place her wedding band, which I have been wearing for safe-keeping, back on her ring finger.

"Oh God, I can't—" she gasps. It is the only time I've ever heard my mother say *I can't*.

"I'm right here," I say, and in that moment, something shifts. A box in my mind opens, and I know that everything I might otherwise feel, think, or say on this day will be placed in that box, deferred for viewing at a later date.

"Hold my hand, Dani," my mother says. "Hold my hand and don't let go."

Hundreds of mourners fill the auditorium of Overlook Hospital. They are standing in clusters, groups of old friends who haven't seen each other in years, business associates of my father's, relatives who hopped planes within hours of hearing the news. The first thing I see, beyond the dark crowd, is a plain pine box in the front of the room. My cousin Mordechai, one of Shirl's sons who is a rabbi, is swaying over the box, holding a prayer book. I see the box and then register that my father is inside it. I've never seen a coffin before, and the realization hits me with all the force of a physical blow. I want to tear the lid off the box, gripped by the insane notion that my father won't be able to breathe in there. I am holding my mother's hand, stealing glances at her. She hasn't seen her husband's casket yet. Who picked it out? It seems awfully plain, as if it will disintegrate quickly—which I guess is the whole point.

Lenny is at my side, and I don't want him there. Suddenly, I can't bear the idea that Lenny is at my father's funeral, but

there's nothing I can do about that now. I keep my body between Lenny and my mother's stretcher. I don't want him
near her.

"Listen," I whisper in his ear, "I have to stay with my mother.
Why don't you go sit over there—" I point to an empty seat next
to a few friends he knows slightly.

He looks at me accusingly.

"Lenny, just do it," I say.

I watch as Lenny squeezes past an older couple and takes a
seat next to some of my college friends. He looks lost in their
midst, like a grown-up who has accidentally wandered over to
the kids' table. He's still glaring at me, and I turn my head away.
Lenny has always told me he would be here for me if the shit
ever hit the fan. *As long as I'm alive, you'll never have to worry
about anything,* he has often said, his voice hoarse with emotion.
As I walk alongside my mother's stretcher as we wheel her down
the center aisle of the auditorium, it occurs to me that Lenny is
here, all right—and he's making things worse.

The orderlies park the stretcher right next to the front row,
near the rest of the immediate family. Susie waves me over, but I
cannot leave my mother's side. Everyone is staring at her. It's hard
not to. Her bed is like a float in a parade, her casts and sheets
ghostly white amid the black suits and dresses of mourning.

Oh, Irene, I'm so sorry—

My God, look at you—

If there's anything—

People descend like vultures. They converge on my mother's
bed, leaning over her, their eyes wet with sorrow. The doctor's
words—*She may not survive this*—swim through my head, and I
gently, or perhaps not so gently, push my mother's friends away.
I see dread on her face. She has seen my father's casket, which is
now right in front of us, and a moan escapes her lips. She
squeezes my hand, the edge of her ring digging into my palm.

The rabbi enters the auditorium, and suddenly everything begins moving too quickly. Time speeds up, and I cannot hold on to it, I can't catch it and slow it down. He begins to speak— this rabbi who married my parents twenty-seven years ago—or perhaps he begins to sing. He has a deep, ringing voice, and he rolls his *r*'s. He seems closer to God than the rest of us. My mouth feels like a cavern, empty all the way to the back of my throat. For a moment, I think that I have been rendered mute, that I may never speak again.

I cannot peel my eyes away from the plain pine casket. It's wider on top, and narrower on the bottom. In fact, the whole thing seems too narrow. How do I know it's my father in that box? What if they made a mistake and switched caskets? As a child, I used to wonder if perhaps my parents got me mixed up at the hospital and brought home the wrong baby. Could it be we're burying the wrong man, and that my father is still alive and ranting up on the sixth floor?

Now my uncle Morton is walking to the podium, holding a few sheets of paper. He looks small and old, with a dark yarmulke perched on his head. Why Morton? He's only a brother-in-law. Why isn't Harvey up there? I steal a glance at my father's brother. He's staring straight ahead, tears rolling down his cheeks, his jaw bunched.

"Is anyone else giving a eulogy?" I lean over and whisper in my mother's ear. This is a detail that, until now, has eluded me. It is possible that I've never used the word *eulogy* in a sentence before in my life.

"John Hirsch," she whispers back. Hirsch is my father's partner, or actually his boss. He's a good fifteen years younger than my father. A portly, rich guy. Come to think of it, sort of a Lenny type.

I don't pause to wonder how my mother arrived at these

choices. She didn't ask me to speak—though God knows, I wouldn't have wanted to. I doubt she asked Susie, or Harvey. I try to focus on what my uncle Morton is saying, but it might as well be in another language. After Morton is finished, John Hirsch gets up and says a few words, words I will not remember. His gold watch, his cracking voice, the feather on the hat of a woman in the third row, and the sound of the rabbi's mellifluous Hebrew as he recites the Twenty-third Psalm—these are what stick to my bones.

In a flash the service is over. The doctors may have asked the rabbi to keep it short for my mother's sake. I noticed several of them standing in the back of the auditorium. Are they there to pay their respects to a man they never knew or to make sure that if anything happens to my mother she receives immediate help?

The rabbi announces that the burial will take place at Washington Cemetery in Bensonhurst, Brooklyn. My Orthodox cousins all seem to know what to do, as if, in the yeshiva, there were classes in the etiquette of death. The men surround my father's casket, hoisting it onto their shoulders, carrying it through the lobby of Overlook Hospital and into the bright February morning. A hearse is waiting by the curb.   ·

In the moments before we leave for the cemetery, I have no idea where I should be, or what I should be doing. My mother is inside the hospital, her family and friends huddled around her stretcher. From what I can gather, they're discussing who should stay with her and who should go to the cemetery. I have walked in and out the sliding glass doors of the hospital a half-dozen times already, watching as my father's casket is loaded into the hearse, running back into the lobby to let my mother know what's going on.

"What about Dani? What's Dani doing?" someone asks, as if I'm not standing right there. A white-hot rage crashes over me, and I turn to see who the asshole is who could possibly imagine that there's a choice in the matter, that I might not go to my father's burial. But something stops me, something stronger than rage; I realize it's guilt that I'm feeling, horrible guilt that I have to choose between my two parents at this moment.

"I'll call you," I whisper to my mother. "From the cemetery. The minute I get there."

She clasps me to her, strokes my hair.

"Oh, Dani, I'm so sorry you have to go through this alone," she says.

Lenny materializes by my side. I had almost forgotten about him in the last half hour. By now my parents have met Lenny. My father has died thinking that this is the life I have chosen.

"She won't be alone," he says, putting an arm around me. I want to shove him away—this anger seems to have taken over my body—but instead I stand there and contort my mouth into a smile.

"Mom, you remember Lenny—"

"Of course." She holds out a hand, like Caesar's wife. "Lenny, so good of you to be here for my daughter."

Somehow, Lenny, my uncle Morton and I end up in the back of the first limo, following the hearse. As a kid, I remember seeing funeral processions on New Jersey highways and asking my father why the limos and cars had their headlights on. He told me it was so that the drivers wouldn't lose one another on their way to the cemetery. I remember thinking how awful it would be, getting lost on the way to the cemetery.

Lenny and Morton are having a conversation about the one thing they have in common: yachts. They each have a yacht—Lenny refers to his as "the boat"—and they are comparing notes, talking across me, about teak finishes and the benefits of fiberglass. Lenny's boat cost over a million dollars. It's sleek and Italian and has everything imaginable built into it. Stereo piped through the walls of the cabins, marble showers, leather couches, and a big-screen television.

"That Riva's quite a machine," says Morton, who is a real sailor, having spent most of his adult life on Hawaii. "You must really let that baby rip."

"I had a forty-two-footer before I got this one," Lenny answers, "and I've got to tell you—"

I fade in and out of their conversation, wavering between numbness and disbelief that they're talking about boats on the way to bury my father. For Morton this is part of grieving, this grasping on to the tangible world—in this case, yachts—because after all, what good would it do to talk about anything else? But Lenny talks about his boat all the time. He could just as easily be on his way to a dinner party. I close my eyes, lean my head back, and try to summon my father's face. I strain to hear his voice, to feel his touch—somehow knowing that these will be the first to fade. Years from now, I will no longer be able to feel my father's hand on the small of my back, or hear the particular way he has always said my name, with traces of a New York accent. But today I still feel him all around me. I pull him closer, like a cloak.

The hearse turns slowly off the Brooklyn-Queens Expressway and onto a wide, bombed-out boulevard in Bensonhurst. Overflowing garbage cans are piled along the sidewalks, wedged into dirty snowbanks. An el train rumbles overhead. I feel as if we're taking my father to some godforsaken place, where he will be lonely.

The funeral procession pulls into the gates of Washington Cemetery, and before I know it, Uncle Harvey has jumped out of his limo and is heading over to a small office. Clearly he knows his way around this place.

"Paperwork," murmurs Morton. "Let him take care of it."

We sit in the car and wait. Lenny and Morton have fallen silent. Yachts have no place here, amid the tombstones. A few minutes pass, and Harvey strides over to us, a muffler wound around his neck and the lower part of his face, even though it's not that cold out. His eyes are glassy.

"There's a problem," he says, addressing Morton.

"What?" I blurt out. I can't imagine, in this context, what could possibly constitute a problem.

"They've opened the wrong grave," Harvey says.

"What do you mean?" I hear my own voice grow shrill.

"Exactly what I said," he answers.

"But how could that happen? Whose—"

"My mother's grave," he says wearily. "They dug up the grave reserved for my mother."

The thin membrane of self-possession that has gotten me through the morning crumbles, and I begin to weep uncontrollably.

"How could they make a *mistake*? How?" I ask, the word stretching into a howl. I am an animal, incapable of thought, blinded by pain.

"Snap out of it, Dani!"

I stare at my uncle. *Snap out of it?*

"So what's happening?" Lenny asks evenly.

"Well—the grave diggers are on their lunch hour."

"When are they due back?"

"I'm not sure. There's some question about whether they'll be able to do it today."

"Have you tried to reason with them?"

Lenny is firing off questions as if interrogating a witness.

"They *said*—" Harvey's voice is getting a familiar edge.

"I don't give a shit what they said. Surely they can be persuaded."

Lenny and Harvey stare each other down, and for a moment I wonder if they're about to get into a fistfight. I feel as if I'm having a nervous breakdown. My body is flying apart—limbs shooting in opposite directions, head twisting off my spine. I look at the hearse, parked at an angle, the back door open, my father's coffin inside. My cousins Mordechai and Henry are standing next to it, *davening*. How long does it take to dig six feet into the ground? I think of my mother, back at the hospital. What am I going to tell her?

Without a word, Lenny hops out of the car.

"Where are you going?" I call after him.

"Be right back," he says through the window.

For the rest of my life, I will know that Lenny Klein paid off the grave diggers, that he reached out a hand with a neatly folded bill just the way he might to a maître d' in a four-star restaurant. Lenny's money got my father buried quicker. The rest of the men—Morton, Harvey, Henry, Mordechai—would never have thought of it. It takes a certain kind of mind to believe that anything can be bought.

Within an hour we are shuffling down the narrow paths of Washington Cemetery, our coats flapping open on this unseasonably warm February day. I am walking with Susie and Shirl. My cousins carry my father's casket on their shoulders. It sways like a small craft on a black sea. The path is slushy, with patches of ice, and the heels of my black shoes sink into the mud.

It happens fast—terribly fast—once we are all gathered around my father's grave. It is the first time I've ever been to the family plot—in fact, it's my first funeral, period. I have no way of knowing that this Orthodox service is harsher than most burials. There is no Astroturf laid over the grave like a bright green carpet. No flowers in sight. Nothing to prevent me from watching as the pine box with my father inside is lowered into the ground with long cloth straps. It tilts violently from one side to the other, and I am afraid it will turn over and my father's body will pitch to the bottom.

Mordechai is singing El Malei Rachamim.

Someone hands me a shovel.

I dig the shovel into the mound of earth, which smells like the gardening gloves my mother used to wear when she planted tulips each fall. I hold the shovel above the hole, and even as I turn it over, I want to stop the dirt from falling, I want to freeze it in midair. I look around me at the stricken faces crowded around the grave like pale bulbs. I feel as if I'm doing violence to my father. The earth falls in slow motion. It hits the top of my father's coffin with a sound like a snowball's dull thud against the side of a house. I double over, holding myself up with the shovel, gasping for air. But then something inside me tells me to go on. Not to stop. To dig deeper, harder, faster.

Others join in, and as more earth fills the grave the thuds grow softer, as if they're farther away. Something about the physical exertion is comforting. After a few minutes Lenny tries to take the shovel from me, but I turn my back and keep going. Finally, I realize I'm shaking and sweating, and the mourners have formed an aisle down which Susie and I are nudged, because the nearest of kin must be the first to leave. I want to stay. There is an old bench in the family plot, and I want to sit there alone, still as stone, and guard over my father's grave until the ground is level, until a hedge has grown above it. I want to spend the night here, listening as the wind whips through the graveyard and the el rumbles overhead. Instead, I do what I have to do. I walk away.

Before I get back into the car with Lenny and Morton, I stop at a pay phone in a gas station across the street from the cemetery and call my mother.

"How was it?" she asks.

A flock of pigeons scatters as a taxi hits a water-filled pot-hole, drenching the side of my black coat. I close my eyes.

"It was beautiful," I say. "It was just what he would have wanted."

CHAPTER
SIX

The night after we bury my father, the skies open up. Lenny and I drive back to the city in the pouring rain. We are listening to a cello concerto on the radio, a mournful piece I don't recognize. My father and I used to listen to classical music and then try to guess the composer. *Saint-Saëns. No way. Brahms. Hmm. Maybe Brahms.* Through the windshield, I can barely read the highway signs or make out the Budweiser plant as we head toward the Lincoln Tunnel. The world is smudged, impressionistic. I keep picturing my father's grave, water seeping through the earth and into his coffin.

We go to La Caravelle for dinner. We have a reservation under *Dr. Klein.* Lenny thinks doctors get better tables. Why we are out at a fancy French restaurant, why we aren't home with my family, is a question I don't think to ask, much less answer. This is what we do, what we have always done, Lenny and I: we go to dinner, get on airplanes, check into hotels for no apparent reason. For a minute, it seems life isn't going to change at all, even though I feel as if someone has come along and spilled out my insides.

Leaving my mother alone in the hospital tonight was almost as difficult as leaving my father alone in his grave this afternoon. I have always imagined my mother will live forever, and now I am convinced that she won't make it through the night. The doctor's words earlier have stayed with me: *Your mother is a very sick woman. She may not survive this.* She has to survive this, she just has to. If my mother dies, she will take with her the only reason I have to live.

Lenny pours me a third glass of Puligny-Montrachet. I have pushed bite-sized pieces of poached salmon around on my plate, hiding them beneath dill sauce and lettuce leaves. I'm slumped against the banquette, trying to summon the energy to lift the wineglass to my lips. I have had enough to drink. The world is at a remove—just far enough away so that I think I can't touch it and it can't touch me. When I was a child, it was discovered that I had a weak eye—and in order to strengthen it, a gray filmy patch was put over my stronger eye so that my vision would correct itself. Now, it is as if I have two of those gray filmy patches over both my eyes, softening the edges of my thoughts. But no matter how much I drink, it doesn't change the facts: my father is dead, my mother may be dying.

The waiter clears my plate.

"Was the salmon not to mademoiselle's liking?"

"No, no, it was delicious," I say, trying to smile.

"Does mademoiselle not have an appetite?"

"Mademoiselle would like a Rémy Martin and the chocolate soufflé," Lenny interrupts.

"But I'm not hungry," I protest.

"You'll eat what you want," he says.

I excuse myself from the table and walk very slowly to the

ladies' room. Blood is rushing to my head, my knees are rubbery, and for a moment I think I might faint. Each step feels
precarious, as if I were on ice, as if I were an old woman and if I
fall my bones will splinter. I'm wearing the black suit I wore to
the cemetery today; I'm supposed to wear it for the entire week
of shiva. The lapel of the suit was torn at the funeral by the
rabbi, directly over my heart. I must be quite a sight, here
among the patrons of La Caravelle. I inhale sharply and smell
dirt.

In the ladies' room there is a pay phone. I dial the number
for Overlook Hospital, which I now know by heart, and ask to be
connected to my mother's room.

"Hello?"

My heart leaps at the sound of her voice.

"Hi, it's me."

"Hi, sweetheart."

She sounds utterly exhausted.

"How are you feeling?"

"Pretty good, pretty good," she says. "Where are you?"

"At dinner with Lenny," I say.

"That's nice. Say hello for me."

I brace myself against the wall and close my eyes.

For as long as I can remember, my mother's energies, at least
regarding me, have been misdirected: intrusive at all the wrong
moments, passive when she shouldn't have been. She has been
stunningly silent about the important things: my dropping out of
college and taking up with Lenny.

My parents met Lenny after we had been together about a
year. He took them to the Harvard Club—even though he didn't
go to Harvard—for drinks, and told them that his intentions
toward me were honorable but his wife was mentally ill and it
would take some time for him to disentangle himself. After all,

there were children involved. As parents themselves, he said, he was sure they'd understand.

I watched as my father scrutinized Lenny. What must he have thought? Whatever my father felt that night—disgust? fear? numbness?—he kept to himself. But my mother smiled encouragingly at Lenny across the table and asked him questions: Where had he gone to law school? (University of Chicago.) How large was his firm? (Two hundred partners.) Where was his house on Martha's Vineyard? (Edgartown.) With each exchange I felt myself shrinking. I could see she was impressed by his credentials, charmed by his suave, flirtatious manner. I glanced at my father, but he was staring at the ice cubes in his drink. At the end of the evening, as the four of us left the Harvard Club, my mother turned to me as we were about to go through the revolving doors: *He's so cute*, she whispered. And whatever small part of me that had held out some notion that my parents might rescue me faded until it was gone.

"After you left, I had some company," my mother is now saying. "The Alenicks stopped by, and Dorothy and Bernie Quentzel. They're coming out of the woodwork. And they're all asking about you, Dani. They all send their love."

"What can I bring you tomorrow morning?" I ask. "I'll be there first thing."

"Nothing, darling. Just you."

"You know we're sitting shiva in the evening," I say. "I should probably order some platters from Fine & Schapiro."

"That's fine."

"So . . . you're going to get a good night's sleep, right?"

"Right-o."

"Mom?"

"Yeah?"

"I'll be here for you. You're not alone," I say, and suddenly the gray patches are gone, and I'm choking back tears. No matter what has gone on between me and my mother, there is a fierce love between us. When I was eight, I spent two weeks in the hospital with pneumonia. My first few nights there I was petrified, and my mother slept on an empty cot in my room. Even though there's no place in my mother's room for me to sleep, I feel that I should be standing guard over her. I should station myself in the doorway of her room and watch her through the night.

I blow my nose, splash cold water on my face, and head back to the table. I can already tell from the look on Lenny's face halfway across the room that something's wrong. He's holding my glass of cognac, and the chocolate soufflé is almost finished.

I slide into the banquette.

Lenny glares at me, his nostrils flaring.

"Where the hell have you been?" he asks.

I stare at him.

"I called my mother," I say.

"Do you realize," he enunciates each word clearly, "that you've been gone *fifteen minutes*?"

"I—"

"Your soufflé is getting cold."

I feel as if someone has peeled back my eyelids. It is impossible to blink. And what I'm seeing is this: a middle-aged man with a trace of chocolate around his upper lip and a linen napkin riding up over his belly, his face flushed like an angry toddler's.

"My soufflé is getting cold?"

"How dare you leave me sitting here like this?" he continues.

I start to laugh. It may well be the most awful sound I've ever made, this laugh. The couple at the next table both turn to look at me. I either have to laugh or take the dull silver butter knife still on the table and ram it into Lenny's heart. I grab my bag and begin to slide out of the banquette, squeezing between tables.

"What do you think you're doing?" Lenny asks.

I smile and keep moving. There is a Hebrew word in my head, *shiva*, which translates into the number seven. I am at the beginning of the seven sacred days of mourning, and to shed even a single tear over Lenny Klein seems like a sin for which I will never be able to forgive myself.

"Where are you going?" I think I hear him call behind me, but stopping is out of the question. I walk through the restaurant, past the bar, out onto West Fifty-fifth Street. It isn't until I am standing on the sidewalk in the torrential downpour that I realize I don't have a penny in my handbag, that all my money is in my coat pocket, and I left my coat in the restaurant. I stand still for a moment, raising my chin to the sky, the rain soaking through my thin suit. It feels as if the world is crying.

Lenny slams out of the restaurant, my coat under his arm. He is breathing heavily, and his eyes are wild.

"That's the last time you ever walk out on me!" he screams.

I gaze at him calmly. There is nothing happening inside my body. No fear, no rage, no regret. I hold out my hand. At first, Lenny must think I'm reaching out to him, but then he realizes it's my coat I'm after.

"How dare you?" he bellows.

He hurls my coat at me.

I catch it, wrap it around myself, and walk away. I hear the melody of the mourner's Kaddish with each step I take as I head for Sixth Avenue, where the white on-duty lights of taxicabs float above the slow-moving traffic. Without turning around, I know that Lenny is right behind me. I can hear his labored breath. When I reach the corner, I stop and wait until he catches up with me.

"I'm sitting shiva for my father, Lenny," I say, looking deeply into his eyes, seeing nothing there.

"Yeah, I'm with you, Fox," he says. "Listen, I'm sorry, I—"

"No, you're not," I interrupt him.

"Yes I am, I'm sorry—"

"No. I don't care if you're sorry. I don't give a shit. That's not what this is about, Lenny."

"What do you mean?"

"For the next seven days, I'm not going to see you, or talk to you. Don't bother trying to call me," I say. "When shiva's over, we'll talk."

He stands on the curb as I step into the street and hail a cab. I know he's waiting for me to change my mind, to turn to him with a shrug and ask him to come home with me. But I don't look back. I think of my father's face the night he met Lenny and the way his eyes watered up as he stared into the distance. If the way he felt about me the last few years of his life could be summed up in a single word, it would most likely be disappointment. All I want in the world is to find a way now to make my father proud.

I am in shul with my father on a bright Shabbos morning. The sun streams through stained-glass windows shaped like the ten commandments. My mother has stayed home; she thinks the rabbi is a buffoon. She usually wakes up with a headache on Saturday mornings and likes to sit quietly in her bathrobe, nursing a cup of coffee. Sometimes I keep her company, other times I go with my father. I love our walks to temple. We take the same route each week, up Revere Drive, across Nottingham Way, down Westminister Avenue, unless it's raining or really cold. In bad weather we take a shortcut across the Pantirers' tennis court

that saves us five minutes. Our neighborhood streets all have
British names, and the houses are a mishmash of New Jersey
architecture: Georgian colonials next to Tudors next to split-
levels. We know everybody—we've lived here since I was nine
months old—and people wave to us as they pass us in their cars.
Sometimes someone slows down to offer us a ride, but we don't
drive on Shabbos.

We've already worked our way through the Orthodox temple
on the other side of town. My mother had a huge fight with the
rabbi of that temple, and it was so bad that we've never gone
there again. So we find ourselves in this bastion of suburban
Conservative Judaism, with its handwoven curtains, wall-to-wall
carpets, and rabbi who makes golf jokes in his sermons.

I am six or seven. Small enough so that my feet don't quite
reach the floor below the wooden bench where we sit. I play
with the fringes of my father's tallis, knotting and unknotting
them. On holidays, he also wears tefillin, small leather boxes
attached to long leather straps wrapped around his arms. He's the
only man in shul who wears tefillin on holidays, the only Orthodox
man. If we were at the other shul, he'd be one of many—but
here he sticks out. I lean my head into his shoulder, close my
eyes, and listen to the murmur of Hebrew as it rises and falls like
the tide. Every once in a while there's a song I know, and I sing it
along with my father, both of our voices loud and off-key.

I don't yet know what the words mean, but they move me,
they transport me to a place where my father and I share our
own language. My mother doesn't have this language. She tries,
but the words come out garbled. Hebrew is something that's my
father's and mine alone. I watch his face as he's singing. He's big
and tall, and has a soft stomach like a pillow. He has sad green
eyes, but when we're in shul his eyes light up. He sways and
smiles and winks at me if he catches me looking at him.

Services end right before lunchtime, and after a quick appearance at the kiddush in the basement, we head home. My father usually wears a hat instead of a yarmulke. I think this is because some of our neighbors don't like Jews, and over the years they've thrown eggs at our house and driven their cars over our lawn. Where my mother loves nothing more than a good battle, my father craves peace. He takes it however he can get it: in the sanctuary of prayer, in the little yellow pills he swallows before lunch.

My mother has set the dining room table formally, with three linen place mats gathered at one end. Cold brisket, cranberry sauce, iceberg lettuce. Her mood is as cold as the lunch she has prepared.

"How was it?" she asks, looking at me.

"Fine," I say in a small voice. I have already noticed her eyes sliding away from my father's. She thinks I'm too young to see what's happening, but I am an only child and I have nothing better to do than to figure out my parents. I know my mother hates it when my father and I go to temple together—though she doesn't hate it enough to join us. I know she gets angry when we do anything without her, though I don't understand why. She needs to be the center of both of our lives, bobbing like a life raft in the middle of a pond, keeping us apart.

There is a loaf of challah next to my father's place, covered with an embroidered cloth. The challah knife was a gift from my grandparents. He says a quick *hamotzi*—the blessing over the bread—as he slices it, scattering crumbs across the linen placemat and onto the floor.

"Goddammit, Paul!" My mother explodes.

"What?"

"How did I marry such a slob?"

"Oh, stop it," he says, and then, in Yiddish, "*der kinder*."

He speaks in Yiddish so I will not understand, but I know *der kinder* means "the child."

I crawl under the table and start picking up the crumbs. There is a small brass buzzer down there, near my mother's foot. She recently installed it so that she can summon the maid during dinner parties without having to get out of her chair. I know I'm not supposed to touch it on Shabbos, and the maid isn't even around today, but I can't help myself. I press the buzzer and wait to be struck down by God. I hold my breath and count to ten—but nothing happens. I press it again.

My father's shoes are shiny like beetles. He has no idea what I am doing under the table; I imagine if he knew it would break his heart. He wants me to be a good little yeshiva girl. I went to temple with him, but now, in secret, I have aligned myself with my mother. I know she smokes cigarettes on Shabbos, that she probably eats *traif* when no one's looking. I stare at her bare ankles, veiny and white above her sneakers, and press the buzzer once more. Something is taking hold inside our house, something so corrosive it eats itself up even as it happens, and leaves no trace.

"Dani, get out from down there," my mother says sharply.

I emerge red-faced from under the table. My father smiles at me and gives my hand a squeeze. His gaze is crinkly and kind, and I feel like a traitor. I already know I can be either my father's daughter or my mother's. I am caught between them in a battle that will last longer than I can possibly imagine. Longer than life itself.

After leaving Lenny standing on the curb in the pouring rain, I cannot imagine going home. The cab splatters him with muddy water as we pull away, and I can see his mouth moving. He's probably cursing me. Where's Lenny going to end up tonight?

Maybe he'll get into his car and drive home to Westchester, where he will tell his wife some fantastic story about how his London business trip got canceled. Maybe he'll check into a hotel and order up a bottle of scotch from room service.

I tell the driver to head toward the Upper West Side. I've recently moved out of the downtown apartment Lenny rented and into a small one-bedroom on Seventy-second Street. I'm afraid to go home. I've barely been in the apartment in weeks. I don't want to face the newspapers piled up outside the door, the spoiled milk in the refrigerator, the dying plants. Besides, what if Lenny shows up? Even though we're in one of our phases where he doesn't have a key, it wouldn't be the first time he's banged on the door, screaming at me to let him in.

"Lady, you know where you're going?" the driver asks.

We are heading uptown on Broadway.

"Give me a minute," I say.

"You've gotta give me a destination," says the driver.

I feel my father all around me, in the gusts of wind so strong the traffic lights sway like pendulums, in the rain pummeling the cab's windshield. I begin to realize that I will never see him again. I hug my knees to my chest.

"Seventy-fifth between West End and Riverside," I say, deciding to ring Annette's buzzer, praying she's home. I have very few options left in my life, almost no one to whom I'm close enough to show up in the middle of the night. Annette and Diane are about it. My other college friends have disappeared— or rather, it is I who have done the disappearing. Not a day goes by without a thought of Jess. She is a constant, ghostlike presence, rising up inside me, shaming me, never letting me forget the terrible choice I made.

We pull up to Annette's brownstone in the middle of the block, and I pay the driver, then ask him to wait while I ring her.

"Yes?" she calls down after a minute, her voice suspicious.

"It's me."

"Who?"

"Dani."

"Well, God, come on up," she says.

Annette lives alone in a studio apartment with two cats. She sleeps on a futon on the floor, surrounded by scripts and Samuel French editions of plays she's studying in acting class. Her black-and-white waitress uniform is flung over the back of a rocking chair, as is the corporate suit she wears for her legal temp job at Lenny's firm. She's trying to make extra money to pay for some decent head shots, which can cost up to a thousand dollars. Her apartment is a comfortable jumble, and I find myself comparing it with mine. My place is spare and weirdly furnished, with the baby grand piano Lenny bought me taking up half the living room, and a big brass bed with white iron bars dominating the bedroom. The closets are stuffed with designer clothes Lenny has bought me over the years, clothes I will one day stuff into giant trash bags and donate to the Salvation Army.

Annette hands me a mug of tea. She's wearing a pink flannel robe, and I want to be her. I want to have a life where robes and cats and mugs of tea are within the realm of possibility.

"Can I spend the night?" I ask.

"Well, it'll be hard for you to stay here," says Annette, "but my downstairs neighbor is in LA for pilot season. I have keys to her place. You can stay there as long as you like."

And so it happens that the night after my father's funeral I am in the apartment of a stranger. Annette leaves me with my mug of tea and an extra blanket, and retires to study her Tennessee Williams monologue for tomorrow night's class. I want to hold on to the sleeve of her robe and beg her not to leave, but instead I hug the blanket to my chest and nod when she asks if

I'll be okay. This is the first moment I have actually been by myself since my father's death. My father is alone in the ground, my mother is alone in her hospital bed, and I am here in the home of someone named Leanne who is in LA for pilot season.

I don't know what to do to quiet my nerves. I pace the floor, roll my shoulders back and forth, shake out my arms and legs like a rag doll. I clear my throat, just to hear the sound of my own voice. I feel as if I've swallowed a grenade and I'm about to explode—in the morning, Annette will come downstairs and find me splattered all over the ceiling. I think about going out to buy a bottle of vodka—there's a liquor store still open on Broadway—or I could beep my friend the cocaine dealer and ask her to make a house call. I'm tempted to do what I've always done, and find a way to short-circuit these feelings instead of actually living with them. But something tells me that what's happening inside me is not about to go away, and there's nothing I can do about it.

Next to the bay window overlooking West Seventy-fifth Street there is an antique rolltop desk and an old wooden chair. I bend over the desk and start riffling through the papers, mail, and magazines lying around. I find an eight-by-ten of Leanne, who, it turns out, is a beautiful redheaded Juilliard graduate with a serious list of credits. She's done everything from regional Ibsen to a small role in a Stallone movie. I picture her now, on the West Coast, where all good little actresses should be. It's ten o'clock in New York, seven in LA. She's probably just finishing up her nightly aerobics class after a day filled with auditions and callbacks.

I pull open the file drawers and begin exploring the life of this woman in whose apartment I am spending my first father-less night. I know I'm doing something wrong, something that would horrify Annette if she could peep through a hole in her floor and see me now, but I can't help myself. I'm looking for a toehold on a slippery slope, and I think I might find it in this desk, as if buried in these drawers might be the secret to living life as a young woman.

For the past few years, I have rummaged through Lenny's briefcase whenever I've had the chance. In the side pocket, he kept one photograph, of his whole family standing in a semi-circle on the beach outside his home in Martha's Vineyard. Lenny is the only man in a crescent of women, and they are all holding hands. I have stared at this picture so often I have it memorized. The wide smile on his wife's face, the way the wind whips her shiny black hair across one high cheekbone. Next to the wife is Jess, beautiful and barefoot in a pair of cutoffs and a tank top. The rest of Lenny's clan—stepdaughters, daughters from his first marriage, and the two little ones he's had with his wife—round out a vision, to my mind, of a perfect family. And in the center of it is Lenny. He's wearing a sweatshirt and jeans, and is smirking at the camera like a man who has figured out the universe.

As I yank Leanne's file drawers open, once again I wonder where Lenny is right now. No matter how I try, I can't get him out of my head. Has he driven back to that semicircle of women? I have often wondered what it must be like to be connected to that many people. I look around this stranger's apartment and think, I could die here and no one would know.

The first time Lenny kissed me I thought I was going to throw up. I had been kissed only by boys at that point—young, handsome boys with firm jaws and soft stubbled cheeks, whose hands roamed my body in fits and starts.

We were in Lenny's car. I hadn't seen him since our dinner at the River Cafe several weeks earlier. The flowers he had sent every single day since that evening, a lost-looking delivery van pulling up outside my dorm for all the world to see—how did he have the nerve? Jess lived on the other side of campus, but she used to hang out in my room all the time. What if she had seen all those flowers? I kept throwing them away, but they kept coming.

At first I didn't tell Jess that I had stumbled into a date with her stepfather because I kept hoping the whole incident would just go away. I thought if I continued to ignore Lenny's advances, eventually he'd grow tired and stop. I would see Jess in the cafeteria for lunch, and we'd sit together by a window and talk about boys and classes and parties. I would admire the elegant angles of her face, her dark flashing eyes, and the way she looked at me as if she knew more about me than I knew about myself. I would feel my cheeks flush, my gaze slide away from hers. I tried not to think about how hurt she would be if she knew the horrible thing I had done.

So when Lenny showed up outside my dorm and sat there in his car until I came out on my way to the library, I panicked. I leaned into the open window.

"What are you doing here?" I asked. "What if Jess sees you?"

He patted the passenger seat next to him.

"Get in," he said.

Glancing quickly around, I jumped inside.

"Lenny, you've got to get out of here," I begged.

"Jess knows," he said quietly. "I told her we went out."

A feeling of unreality came over me. I couldn't imagine I had heard him right.

"What do you mean?"

"Jess knows," he repeated. "I didn't want there to be any secrets."

I was stunned, speechless as we drove off campus. I didn't try to stop him or get out of the car. Something inside me had begun to shift and crumble. When had he told her? What had he said? Was that why she looked at me the way she did, as if we shared some sort of unnamable secret?

Lenny and I drove upstate. What I remember is in fragments, ragged bits and pieces left of an old puzzle: the way the wind whistled through my ears in the convertible, a jar of strawberry preserves he insisted on buying me at a country café somewhere in Westchester, sunlight splaying across a wooden table as he leaned forward and seemed to focus his attention, laser-like, on me. He never stopped telling me how beautiful I was, how special I was, how he thought from the moment he first saw me that I was the most perfect creature he had ever laid eyes on. *I told Jess then,* he said. *I said, that new little friend of yours, she looks like an angel.* He told me about an old girlfriend of his, the actress Jacqueline Bisset, and how much I reminded him of her. His words had a narcotic effect. I thought of Jess, back at Sarah Lawrence, and couldn't imagine facing her, ever again. College seemed light-years away. I became sleepy. I unfurled and stretched out like a cat in the sun.

And when he pulled the car to the side of the road after lunch and kissed me, his tongue pushing deeply, immediately, into my mouth, at first I felt sick, but then I began to float away. I put my hand on the nape of his neck and felt a strange edge there, a piece of hairweave, stiff as a carpet. *You've done it now,* I thought to myself. Somehow it seemed impossible to go back, to stop. I felt myself spiraling downhill, to a place where friendship and honesty and loyalty did not exist.

When I think of anything that's ever harmed me—cigarettes, alcohol, cocaine, Lenny—they've all had one thing in common: the revulsion, the nausea that I've had to fight past before I could take them in. Lenny's tongue in my mouth that day felt intensely similar to the first time I took a deep drag on a cigarette and felt smoke choking my lungs, or the way scotch tasted as it burned its way down my throat. As with all the rest, I told myself that I could handle it—that I *had* to handle it. It seemed to me, at the age of twenty, that I had already ruined myself.

I spent the night after my father's death thumbing through the file drawers of a young woman who seemed to be living a healthy, normal life. Carefully tucked into hanging folders were her xeroxed receipts, expense reports, correspondence with her agent. There was a manila envelope addressed to her from Ohio—three pairs of panty hose sent to her by her mother, along with a few clippings of interest from her hometown paper. I gazed longingly at her eight-by-ten and she smiled back at me, bright-eyed and confident.

I shudder at the memory of myself, hunched over her personal papers, letters from ex-boyfriends, family photos, tax returns. I desperately wanted to trade places with her. I thought of what she'd see if she looked through *my* desk drawers: unopened bills, undeposited residual checks, angry letters from Lenny, tiny jars of cocaine, an expired credit card I use to chop it up. In years to come, I will occasionally catch a glimpse of a familiar-looking redhead in made-for-television movies and realize where I know her from. I never closed my eyes that night. The rain stopped pounding against the side of the brownstone at some point, and the city was as still as I've ever heard it. And when the pale, thin light of dawn trickled through the bay window, I rose from the desk and looked out onto the deserted street. I wasn't even sure I still existed. The whole world seemed empty, for a moment washed clean.

My mother is on the phone when I walk into her room. I'm carrying an enormous platter of cold cuts from Fine & Schapiro.

I didn't know what to get, so I got a little of everything: corned beef, pastrami, tongue, turkey, potato salad, coleslaw. Her eyes widen when she sees it all. There's another platter in the trunk of my car, along with some big bottles of seltzer and Diet Coke, which I'll bring up to her room later. People will be coming to sit shiva, and when Jews get together, food is important.

The blinds have been tilted open, and stripes of sunlight fall across the sheets and casts. There is a breakfast tray pushed to the side, with a half-eaten plate of eggs and a full glass of orange juice still on it. Angie is in the corner, knitting, and the television hanging from the ceiling is broadcasting a midmorning talk show.

This is your life, I think to myself as I look around the room. *This is home, for now.* I rest the platter on the windowsill, then remove a vase of dead gladiolas. The smell nearly makes me retch. The heating vent is next to the window, and it's cooking all the flowers into stiff arrangements that crumble to the touch. There's a basket on the floor, filled with get-well cards addressed to both my parents and condolence cards addressed to my mother.

She sees me looking at them and puts her hand over the receiver.

"We'll have to order those cards you're supposed to send," she whispers.

I nod, and make a mental note to find out what cards she's talking about. This morning I went to a bookstore and browsed through the shelves, looking for something, anything, to guide me through this process. I thumbed through *When Bad Things Happen to Good People.* I looked at a bunch of books with death in the title, until finally I settled on *The Jewish Way in Death and Mourning,* a book that describes in great detail the way the body is prepared for Orthodox burial, the prayers that are said before, during, and after services, and the ritual of sitting shiva. Nowhere did it say anything about sending cards.

She hangs up the phone with a sigh.

"That was your father's banker," she says. "He called to say he's very sorry, he saw the obit in the *Times,* and by the way he's sealing the vault."

"What does that mean?"

"The banks have to do that when someone dies, until the will has been probated," my mother says.

Wills, vaults, probate. Words I've never used before. Made-for-television words uttered by actors playing lawyers. I feel that familiar get-me-out-of-here feeling, but there's nowhere to go. I sit on the edge of my mother's bed. She looks thin. I offer her the glass of orange juice, but she shakes her head and pushes it away.

The phone rings, and she doesn't move to answer it. She is inert, staring straight up at the ceiling. Angie and I raise our eyebrows at each other. My mother seems flattened, as if the banker's call has forced reality to seep in—just in case things weren't real enough.

"Would you get it?" she asks on the third ring. "Just say I'm not here."

Is she joking? My mother has never had much of a sense of humor. Or, at least, what's funny to her never seems funny to me.

I pick up the phone.

"Irene Shapiro's room."

"Dan?"

"Susie?"

"Yeah, hi. Listen, I'm between patients, but I just want to let you know we're sitting shiva at Gram's tonight."

"What do you mean?"

"It just seems easier, with everyone already in the city," Susie says. "Shirl, Moe, and Harv—well, it's asking a lot for them to make the trip out to Jersey."

"Oh," I say. "Oh, okay."

My mother scans my face.

"Whose decision was this?" I ask.

"What?"

"Whose—"

"Oops, that's my buzzer," Susie says. "Gotta go."

She hangs up, and I look at the clock. It's precisely fifteen minutes past the hour. I picture Susie opening the door to her office and ushering in her ten-fifteen patient.

"That was Susie?" my mother asks.

"It sure was," I reply, trying to keep my voice from giving anything away.

"What did she want?"

How do I tell my mother that shiva for her husband is going to take place in an apartment on West Fifty-seventh Street, many miles from here—that it appears my father's family is abandoning her?

"Well, she called to tell us they're sitting shiva at Gram's," I say slowly.

"I don't understand."

"The Shapiros are going to sit shiva at Gram's apartment," I repeat.

"Gram's?"

My mother's incredulity probably stems from the fact that my paternal grandmother has been comatose for the past two decades. They can't possibly be doing it for her sake—she doesn't even know her son is dead. My mother looks so small and lonely beneath her sheets, I would do anything to save her this.

"I can't believe it," she says, and her chin juts into an expression I haven't seen once since the accident. "How dare they?"

"It doesn't matter—" I try to soothe her, but she interrupts me.

"Let me see the paper," she says.

I hand her a folded copy of *The New York Times* and she flips through it until she finds what she's looking for.

"Hand me my glasses," she says, not looking up.

Once her bifocals are perched on her nose, she runs a finger down the small print of the paid obituaries.

"Here it is," she mutters. "I knew it, I just *knew* it. Damn them!"

"What?"

"Take a look for yourself."

" 'Paul Henry Shapiro,' blah, blah, blah"—I read out loud—" 'shiva will be held at the home of Beatrice Shapiro, 345 West Fifty-seventh Street—' "

"Don't you see?" my mother asks.

"What are you talking about?"

The edge in her voice, the way her pupils jiggle almost imperceptibly behind her bifocals—it's all too familiar to me. I have been afraid of my mother's temper all my life. When she flies into a rage, she will say anything to anybody. She once called my childhood dentist a pig in front of his whole waiting room because he was late for an appointment.

"They must have phoned this in yesterday," she says. "They've known all along."

I can barely bring myself to think about the fact that my father's family is abandoning not only my mother but me. Though they're probably not even aware of it, they are forcing me to make an impossible choice: sit shiva for my father or stay here with my mother.

As my mother's daughter, I have relinquished my rights to be a part of their world. I am as foreign to my father's family as they are to me. My mother always ridiculed their lives as small and pious, and made sure we didn't have anything to do with them. All these years, I never realized that I might be missing something.

But now that my father is dead, I see that they are the invisible thread that connects me to him. I need them, and they are nowhere to be found.

It is the last day of shiva—almost a month since the accident— and my mother's face is beginning to look normal again. The swelling has gone down, and the bruising around her nose and jawline has faded and become mushy and banana-colored. She's looking altogether better, and whenever anyone comments on this, she attributes her speedy healing to good genes.

"My side of the family heals well," she tells me proudly. "Your father's family scars like nothing you've ever seen, but *we're* survivors. We snap right back as if nothing's happened."

She tells me this as if I were not also my father's daughter, when the fact is, I have inherited that particular trait from him. I am not as physically strong as my mother, and I don't bounce back as if nothing's happened. If I had been in the passenger seat of the Audi that night, I doubt I would have lived.

But I tell my mother none of this. Instead, I unpack today's bag of gourmet goodies I brought from the city. I'm trying everything I can to inspire my mother to eat: raspberries, crème fraîche, brownies, cheddar quiche. There's a tiny specialty store in my neighborhood, and each morning, before I leave for the hospital, I double-park my car on Columbus Avenue and run inside to see what looks good.

"Mmm, berries," my mother murmurs.

"There are fresh scones, too," I say.

My mother turns to Angie, who is sitting in a chair between the bed and the window, watching the muted television.

"Angie, look at how my daughter takes care of me."

Angie smiles at me, and I wonder what she really thinks.

She's seen it all: my father, my mother's family, Lenny, and the way I come back after lunch each day smelling like vodka and Velamints. I am taking care of my mother, it's true, but I am not taking care of myself. Does Angie worry about my driving back to the city late at night with a few vodkas under my belt, the car window cracked open, flicking my cigarette ash into the darkness?

I haven't spoken to Lenny since I left him on the street, and I've shown up at my mother's room each morning to spend every minute of shiva with her—but I haven't been able to stop drinking. Every morning I wake up and tell myself I won't drink, but when lunchtime rolls around and I drive into Summit to take a break, the pressure in my head grows so unbearable that it seems only a drink will make it go away. And it does—for a little while—until it returns with a vengeance.

"Listen, Dani."

My mother's voice has that I-have-something-important-to-tell-you-and-you're-not-going-to-like-it edge.

"Susie and Shirl want to go out to Tewksbury."

"Yeah, so?"

"Well, they need papers of Dad's. There's some Shapiro Foundation business, and some insurance policies Shirl seems to feel only she can recognize—"

Her voice falters.

"And?"

"I want you to go with them."

"Why?"

"Let's just leave it at that. I want you to go with them, and be with them while they're going through Dad's things."

"You want me to *spy* on them?"

My mother looks at me, stricken.

"No! I wouldn't put it that way."

"You wouldn't put it that way, but that's what you're saying, isn't it?"

"Dani, there are some things you just don't understand," she says.

"And when is this stealth mission supposed to happen?"

"Tomorrow."

I make a bargain with myself. I will do my mother's bidding, and keep an eye on my half sister and aunt as they go through my father's things. But in exchange for this, I will spend tonight in the city with my father's family. I will sit on low benches with them surrounded by fruit baskets and smoked-fish platters, and I will think of my mother, alone in her hospital room. No matter which way I turn, I am a traitor. Tomorrow I will betray Susie and Shirl. Tonight I will betray my mother.

The following afternoon, when I get to Tewksbury, Shirl is there, waiting for me. She tells me Susie isn't coming; she couldn't reschedule her afternoon patients. At first I'm relieved, but then I realize this probably means I'm going to have to make another trip out here to spy on Susie separately. It doesn't dawn on me that I can say no to my mother—that I don't have to take part in what seems like paranoia. What, exactly, is she afraid of? I know my mother doesn't trust my father's brother, Harvey—but Shirl has always struck me as good and decent. Even though she and my mother have kept an icy distance over the years, I cannot imagine Shirl doing anything to hurt her brother's wife. I fiddle with all sorts of unfamiliar keys until finally I get the alarm turned off and the back door open. Shirl has never been here, and it's only my second trip. It almost feels as if we're breaking into a stranger's house. (Many months from now, it will in fact be a stranger's house. My mother will sell it, and movers will come in and pack up all my parents' art, furniture, books, and clothing and put everything into storage. My mother will never see this house again.)

"Dani," Shirl's voice lifts and falls over my name and turns it into something musical. "Why don't we have a cup of tea."

I put up some water to boil in the kettle. I rummage through drawers, looking for tea bags, coming across evidence of each of my parents' idiosyncrasies: my father's papaya enzyme tablets, hundreds of my mother's yellowed clippings of Jane Brody's pieces on health in *The New York Times*. The house seems alive, waiting for them to come home. Shirl sits at the kitchen table, staring out the back window at the desolate backyard, the dark-green pool cover sagging with the weight of melted snow. She looks so sad that I come over and give her a kiss on the cheek.

"I can't believe he's gone," she says.

I swallow hard, fighting back tears.

Shirl grasps both my hands.

"I told your father I would take care of you," she says. "I promised him I'd never lose track of you."

"Of course, of course not," I say. But I am thinking of the few times I have seen Shirl or her family over the years, usually at huge family weddings or Bar Mitzvahs. We visited them in Boston once, when I was a little girl.

"It isn't going to be easy, Dani, but I really want us to be friends. There have been a lot of misunderstandings over the years. Your mother . . ." She trails off, shaking her head sorrowfully.

Shirl looks at me then, and I realize that she knows. She knows why I'm here, that I am my mother's agent, sent here to make sure that she doesn't steal one of my father's cashmere sweaters or lift the de Kooning off the wall.

"Come on. Let's go upstairs and get this over with," she says, giving my hand a little squeeze. And as she follows me up the curved staircase to the second floor of my parents' house, past the blown-up head shot of my tousled, smiling self, I tell myself that it won't be a problem. I don't have to take sides. I've lost my father, and that's enough. I don't have to lose the rest of his family.

Little do I know that it's only just beginning. The bad blood that has existed between my mother and my father's family all these years is seeping through the generations. In his absence, it spreads like a stain. There's no place to run. I can only stand here, helpless, watching as it covers me.

I remember that fall day when I returned to Sarah Lawrence after my drive with Lenny, I felt jelly-legged, drugged. He winked as I got out of the car and said he'd call me soon, and I had a pretty good idea that for Lenny, soon meant hours, if not minutes. I felt like a different person as I climbed the stairs to my room. I would forever be a girl who had kissed the married stepfather of her best friend on the side of a Westchester road. I could never undo that.

There was a folded note pinned to my door, my name written in the bold, black strokes of Jess's fountain pen; I took a deep breath as I grabbed it. *Meet me at the pub at seven.* It was unsigned.

I checked my watch. It was nearly seven. I had been out with Lenny since noon. My world was completely turned upside down, and I didn't know what to do. My cheeks were scraped raw by his thick stubble, and my mouth felt swollen. I wanted to run away. For the first time since leaving home to go to college, I wanted to call my parents and ask them to come pick me up and take me back to New Jersey. On some level, I knew I was in way over my head, that the events of the previous weeks were beyond my understanding. But I was twenty, and I felt too old to ask for help. I had made a mess of things, and now I had to pay the price.

I walked to the pub. Through the window, I could see Jess at our usual corner table. She was leaning on an elbow, her hair practically falling into her beer. I tried to read her face, the way she was holding herself, but she seemed purposely blank, affectless.

"Hi," I said softly, sliding across the table from her.

She just sat there and looked at me, her usual little smile playing around the corners of her mouth.

I felt like sliding through the floor.

"Say something," I pleaded. "Anything."

She shook her head slowly from side to side.

"I'm so sorry," I said. "I never meant for this to happen."

Silence.

"He tricked me," I went on. "I met him because I thought he wanted to talk about *you*."

"Have you done anything with him?"

Her voice was uncharacteristically shrill, and it was the only thing that gave her away. Other than that voice, she was the picture of languor.

I squeezed my eyes shut.

"Today," I answered, "I kissed him."

She pushed back her chair from the table, and I was afraid she was going to leave.

"Jess, he disgusted me," I said quickly. "I was completely grossed out."

She stopped, that familiar, amused glint returning to her eyes.

"Really," she said.

"I swear to you, it will never happen again."

*   *   *

In the weeks that followed, I kept my word to Jess. I didn't answer my phone, and I left my dorm through the back entrance in case Lenny was spying on me. I told no one—not my other friends, certainly not my parents. Jess allowed me back into her life; we never talked about it. It was as if it had never happened.

Then one afternoon Jess invited me to her twenty-first birthday party, which was going to be held at the Westchester home of her mother and Lenny.

"Oh, Jess, I can't possibly go," I told her.

"Why not?"

"I don't think I should see Lenny."

"Lenny won't be there," she replied. "He's going to be away on business."

The night of Jess's birthday party I drove alone to the estate in upper Westchester. I pulled into the long driveway, parked, and made my way up the front walk to the warmly lit white-brick house that, despite my having grown up in a neighborhood of beautiful homes, impressed me with its sprawling elegance. I peeked through the open front door, and saw Lenny standing there, his back to me.

I could have run back to my car. No one had seen me. But instead, shivering, I pushed the door open and walked inside. Jess rushed over to me, drink in hand. She looked particularly beautiful, her hair gleaming in the candlelit foyer.

"Try this," she said, handing me the glass. "It's a negroni."

"What are you doing?" I whispered. "You told me he wouldn't be here."

She smiled at me and wafted away, leaving me standing there.

I got drunker that night than I ever had before. The negroni tasted delicious; it was a pinkish-red drink, and I didn't know it consisted of three different kinds of alcohol. I quickly downed the first one, then asked the bartender for another. I kept circling the room, staying at a distance from Lenny. I knew he was staring at me, but I wouldn't look back. I drifted in and out of conversations with other guests but all I could think about was Jess. Clearly, she had tricked me into coming to this party. She had known Lenny would be here all along. But why? I felt like a pawn in some elaborate game, and it infuriated me.

Close to midnight, Lenny gathered everyone together and made an announcement.

"If you'll all follow me," he said, "I have something to show you."

The whole crowd moved outside, around the back of the house, to the former horse stable that had been converted into a four-car garage. With a flourish, Lenny opened the nearest garage door. A shiny black Alfa Romeo was parked inside.

"For the incomparable Jessica, on her twenty-first birthday," he announced. Jess clapped her hands and laughed with delight.

She looked at me then, out of the corner of her eye. What did she want? Was she testing me? I felt like slapping her.

Lenny caught up with me as everyone trooped back into the house. It was the first time he and I had spoken since he dropped me off at Sarah Lawrence that afternoon almost a month earlier. His wife was walking just ahead of us, in a group. She was even more beautiful than Jess, and seemed, at forty, young and glamorous.

He bent down and whispered in my ear.

"What do you have on under that dress?"

I felt my legs wobble, a wave of heat. In my drunken, confused state, Lenny was suddenly appealing. He was Lord of the Manor, bestower of Alfa Romeos, a benign patriarchal figure.

"If you call me, I will call you back," I said, turning to him. In the dark of the night, I saw his eyes shining, and a flash of white teeth.

"Well now, that's good news," he replied.

That night, I was too drunk to drive back to school, and Jess insisted I sleep over. I told her we had to talk, but I was slurring my words, and when I tried to focus my eyes, I felt nauseated. She made up the bed in the family room, and as soon as my head hit the pillow, I passed out. I dreamt terrible dreams—dreams unlike any I'd had before or since. I dreamt Lenny Klein was standing over me, legs spread, a steady stream of urine splashing over my face. When I awoke with a start, he was walking through the family room in the dim shadows of dawn, wearing a suit and carrying a briefcase. I pretended to be asleep until I was sure he was gone.

CHAPTER
SEVEN

Sheldon is sending me out on auditions again. He's an agent who just won't quit. He decided a long time ago that he was going to make me a star, even though I haven't exactly cooperated. My last job was months before the crash. (I have begun to think of my life as b.c. and a.c.—before crash and after crash.) It was a Coke commercial broadcast during the Superbowl. In it, I play the girlfriend of a guy who has just gotten his driver's license. I toss him a Coke, and there's a close-up of me leaning my head back, guzzling.

I seem to have lost whatever small spark I had to begin with. I can tell that casting directors are disappointed the minute they look at my face on their video monitors. I'm about to turn twenty-four, and I can no longer play teenagers. There's something behind my eyes now, a shadow that will never go away. A film actress might be able to get away with this kind of darkness, but not me. Sheldon is starting to get feedback about my "low energy." I don't know what to tell him. It seems the life has gone out of me. I can't bounce back from the booze anymore. It's showing in my face.

He calls me one morning, and starts talking excitedly into my machine. He knows I'm there, and that if he talks long enough I'll probably pick up. It's a good thing he can't see me. I'm sitting in a rocking chair in a ratty blue bathrobe, pressing wet tea bags on my eyelids. Without Lenny around to distract and torture me, I've been crying myself to sleep every night. I'm on my fifth cigarette of the day.

"Dani, listen. I've got something fabulous. You wanted serious? You got it, babe. Jay Binder is casting the new Horton Foote play. Off-Broadway. Great playwright. And get this: they want an actress who can play classical piano. *Really*."

I pick up the phone.

"Hah! I knew that would get you," Sheldon says.

"When's the audition?" I ask.

"Well, that's the thing. There's a piece of music they want you to learn. Chopin's something-or-other. Where is it—" I hear him flipping through some papers. "Yeah, here it is—the Revolutionary Étude. Know it?"

I laugh.

"Sheldon, it's incredibly difficult," I say.

"Well, just learn the first page," he says.

"By when?"

"Tomorrow, one o'clock."

I sit down at the piano for the first time in at least a year. Stashed inside my piano bench are the complete Chopin Études, because when I played, I was in the habit of buying pieces of music that were over my head. I look down at my fingernails—chipped red ovals—and remind myself to clip them before tomorrow. As I stare at what seems an impossible piece of music, I tell myself I can do it. For some reason, it seems a worthy challenge.

I begin to pick my way through the first bars of the Revolutionary Étude, which consist of a series of strong treble chords followed by great rushes of scales. I think of Mr. Lipsky, my old piano teacher. Thank God he can't hear me butchering this. But I focus—it is perhaps the first time I've focused on anything outside of my family for months—and before I know it, hours have passed, and the western sun is shining through my windows.

The following afternoon, I show up at a small theater in the West Forties. The casting director is there, and the director, and the playwright. As usual, there's an assistant sitting slumped on a metal folding chair onstage; it's his job to read with me—he's been reading with actress after actress all day. By now, he's tossing back his lines in a bored monotone, not even looking up from his script. I struggle my way through two scenes I've only had a chance to skim. It turns out I'm supposed to play a small-town Southern girl who discovers she has musical talent. I know I can't do a Southern accent to save my life. By the time they ask me to sit down at the piano, I'm drenched with perspiration.

My back is to them. I arrange the sheet music in front of me, then take a deep breath and strike the first chord. My fingers have memorized the broken scales that follow, and they fly over the keyboard until they fly right off the piano. Five bars into the Revolutionary Étude, I have run out of keys.

I turn around, confused.

"What's the matter, love?" one of them calls out from the darkness of the seats below the stage.

"This isn't a piano," I say.

"No, actually, it's a piano*forte*," the voice says.

"But—there aren't enough keys. I can't keep going."

"That's all right, love. Thanks for coming by."

*   *   *

Maybe dropping out of college wasn't such a good idea. I walk out of the theater and find myself wandering slowly downtown. I am dressed in my usual I'm-an-actress garb. Something short, something young. I walk past a construction site, and a group of workers sitting on the sidewalk during their lunch break whistle and make sucking sounds. I lower my head and walk faster. *Whatsa matter, baby, can't you smile?*

I head south on Broadway, jostled by the lunchtime and matinee crowds. There's still time to visit my mother, but it's two months since the accident, and I'm trying not to go out to the hospital every single day. I'm trying to get on with my life, as people have suggested. Problem is, I don't quite know what life I'm supposed to be getting on with. I list my attributes in my head: I'm a college dropout, a joke of an actress, an ex-mistress, and I have no money in my checking account. For all I know, I'm in big trouble: I've never paid taxes, barely pay bills, don't balance my checkbook. No one has ever told me these are more than annoying little tasks but are, in fact, a necessary part of living. I have skipped a developmental stage—late adolescence, early adulthood—during which most people learn these survival skills. I've made some of my own money, but my years with Lenny have thrown my sense of financial reality out the window. I know I'm supposed to be getting a check sometime soon from the New York Stock Exchange: a $25,000 stipend to the children of members who have died.

I walk until the sky begins to fade, and I find myself on the cobblestoned streets of the West Village, near the place Lenny and I had a couple of years ago. I plop myself on a stoop and feel through my jacket pockets for a cigarette and matches. Some kids are playing kickball in the street. The ball rolls near my foot, and I kick it back to them.

What's Lenny doing right now? Canvassing the ranks of the under-twenty-five set for a new girlfriend? I blow smoke rings into the air, and watch as they fade to nothing. I feel as if I've been living under a rock for these past four years and life has gone on without me. There isn't a part of this city that I don't associate with Lenny. I have to reduce it all to rubble in my mind and build something from scratch. I don't know where to begin.

Most days now, when I show up at the hospital, my mother already has guests. Men with slicked-back hair and shiny briefcases sit by her bedside, taking notes. The lawyers (there are now a battery of lawyers: malpractice, insurance, trusts and estates) call her Mrs. Shapiro. They call me Dani. I don't trust them, and they know it. I think they're all ambulance chasers, that they're here to make some money off my mother. But she clearly wants them here. She has summoned them to her throne. As far as my mother is concerned, my father's doctors screwed up; they should have seen this coming. That or his brokerage firm should pay damages, since he was on his way home from a business trip when they had the accident. And if all else fails, there's always the possibility of a class action suit against Audi for faulty brakes. My father has died, and it seems she fully expects someone—anyone—to pay for it.

My mother is on a roll. From her bed, she has turned into a miniconglomerate. Her phone rings nonstop, and the bouquets of flowers on her windowsill have been replaced by piles of papers and folders. She is being propelled through the weeks and months by a terrible rage. This rage does my mother good. It brings a flush to her cheeks and makes her sit up straighter. It is the fuel in her blood, the fire that welds her bones back together. The crash has made her tough, invincible. She has survived eighty broken bones and the loss of her husband.

"Guess what?" she says one morning as I walked through the

door of her room, carrying a box of gourmet pizza. I know by the way her pupils are jiggling that something's up.

"What?" I respond, lowering the bags to the floor.

"Your sister is contesting your father's will."

This is one of those sentences I imagined I would live my whole life without hearing. For a moment, it does not compute.

"What do you mean?"

"Look at this."

My mother hands me a letter on a law firm's stationery. I skim it, my mind buzzing with a feeling I don't have a name for. I can barely focus on what the letter says. The words swim on the page.

"Your sister," my mother says, omitting the *half*, "has no idea what she's doing."

"When did you get this?"

"This morning."

My mother's voice is ice. She is drumming her fingers against the rail of her bed. I look over at Angie, who is sitting in her usual spot in the corner, knitting. Angie's house must be filled with needlepoint pillows and intricately knitted throws. I imagine she plops down on her couch every night and thanks God she isn't part of our family.

"It seems your sister doesn't think the terms of your father's will are *fair*."

"Well, are they?"

My mother glares at me, as if I should have known better than to ask such a question.

"Of course they are. You know your father—he was meticulous."

It is precisely because I know my father that I asked the question, but I decide not to pursue this line of reasoning. There's no point arguing with my mother; there never has been.

"So what happens now?"

"What happens now?" my mother repeats shrilly. "What happens now? Now, the will will be held up for God knows how long, which means the estate's assets will be frozen."

"What does Susie want?" I ask.

"God knows what she wants."

I have begun to notice that my mother often invokes God when she's in a rage. In her anger, she believes she's on the side of the righteous.

"Susie doesn't like a certain phrase in the will."

"What's the phrase?"

"As you know, Daddy left half of everything outright to me, and half in trust to you and Susie, which you get when I die."

"Right," I say hollowly. Whenever we start talking about legal specifics, my mind goes numb, as if I'm missing a certain gene my mother and sister both clearly possess.

"There's something in there, just legal language really, which says that I'm allowed to invade that trust five percent a year."

Even I can do the math on this. If my mother takes five percent a year out of the trust, in twenty years it will be seriously depleted. No wonder Susie's upset. It's safe to say she's being screwed. Like everything else in his life, my meticulous father managed not to think this through. The man who imagined his own death every minute of every day left a ticking time bomb of a will that could do nothing but rip his family apart.

". . . did you hear me?"

My mother's voice comes at me from a distance, and I realize I've been spacing out.

"Sorry, Mom."

"I asked if you think you can talk to your sister."

"You mean talk her out of it?"

She looks at me.

"Never mind," my mother mutters. "I don't want to involve you. This is terrible."

She sits up straighter in her bed and tosses her head back defiantly.

"Hand me the phone, honey, would you?" she asks.

I begin to take the pizza out of its box, trying to block out my mother's voice as she gets one of her lawyers on the phone. *My stepdaughter*, I hear, and *frozen assets*, and *countersue*. I'd like to rail against my father, who should have foreseen this, but he's dead. I'd like to talk sense into my mother but I just don't have what it takes to reach her. And while I understand why Susie's doing what she's doing, still I wish she wouldn't do it.

Though I don't know this today, I am making a choice. I am choosing not to choose. My mind has gone numb precisely so that I won't understand that somewhere buried deep in this mess there's a right and a wrong. Someone is the villain here. I just have no idea who it is, and I don't want to know—because knowing will mean I will have to choose my mother over my sister, or my sister over my mother. I've already lost so much I can't bring myself to lose any more. For the rest of my life, I will become stupid around this subject. I will shrug my shoulders when people ask me why my sister and mother don't speak. *My sister contested my father's will*, I will say, as if this is the end of the story—or the beginning.

It's a Saturday night, and my friend Diane and I are out at a jazz club—a plush, dimly lit place dotted with purple sofas and mirrored cocktail tables. We're in the dead center of the 1980s, and clubs like this are thriving. At the table next to us, there are three swarthy guys in suits, their faces flecked with light cast from a mirrored ball slowly revolving on the ceiling; they keep trying to send us drinks, but the last thing we want to do is meet men. I have not spoken with Lenny in two months now, and as far as I'm concerned, men are nothing but heartache. They lie to you, they leave you, or they die.

Diane has ordered a bottle of champagne, not because we have anything to celebrate, but because champagne is her drink of choice. We don't stop to consider the fact that we don't have real jobs, and that sixty-dollar bottles of champagne are perhaps a bit excessive. I have been taking my American Express bills as they arrive each month and shoving them into the bottom drawer of my desk. I receive messages on my answering machine on a regular basis now, demanding that I call the toll-free numbers of collection agencies.

Diane clinks her glass against mine and grins at me as if we're partners in crime and we've just gotten away with something. She is my closest friend now; we are bound by the fact that we have both lost a parent. No one else understands. That she is also a girl who likes to drink—that she can, in fact, drink me under the table—is perhaps an even greater reason to hang out. We get together a couple of times a week, always choosing pretty places where the wineglasses are big and the waiters don't blink when we order a second bottle. We pretend we are doing something other than getting shit-faced. We chain-smoke Marlboros and order only dessert. By the end of these nights, our table is usually littered with lipstick-stained napkins, overflowing ashtrays, empty bottles, and half-finished crème brûlée.

She presses a tiny envelope into my hand under the table. I've sworn off cocaine for the past couple of months because I know I can't handle it, but tonight my resolve has weakened. Champagne, soft jazz, caviar spooned onto toast points—none of it is taking away the dull, empty ache I seem to always carry with me these days, but maybe a few lines of blow will do the trick.

I go down a flight of carpeted stairs and find the ladies' room, which is the size of my whole apartment. The lights in the bathroom are dim as well, and there is a tray with an assortment of hair sprays, colognes, and Q-tips on a long marble vanity. I enter a stall, do a few quick lines off the metal toilet-paper dispenser, then check my face in the mirror. I look pale. I pull a dark red lipstick from my bag, and try to apply it, but it's like putting on lipstick after visiting the dentist. I can barely feel my lips.

I head back upstairs, move quickly past the table of men, and settle back into the purple banquette with Diane. She has ordered another bottle of Cristal. The band starts a second set after a long break—it's one in the morning—and I notice that the sax player is cute. He's a skinny, dark-haired guy, and I find myself wondering what his life is like. He catches me looking at him, and winks. I flush and look away. I have been playing the part of the femme fatale for so long now that I have all but forgotten my true nature. I'm a good girl at heart—a girl for whom rebellion once constituted driving on Shabbos. The fact that I'm sitting here in a jazz club at one in the morning, stoned and drunk, flirting with the sax player—it's as if somewhere along the way there was a fork in the road and it would be an understatement to say I took the wrong one. Retracing my steps back to where I started seems impossible.

"I want some more cookies," I whisper.

Cookies is our code for blow. We have reduced it to something sweet and wholesome in our minds. I pocket Diane's little envelope again and head back to the ladies' room. That's the thing about cocaine: I never want just a few lines. Once it's in my system, I crave more. There is no enough. I turn to smile at the sax player as I leave the floor and catch him staring at my legs. What does he see? On nights like this, I feel transparent. People can see right through me. My real life is in the hospital with my mother. There, I am flesh and bone.

I pour twice as much coke onto the toilet-paper dispenser as I did the first time, cutting four fat lines with the edge of my American Express card. I snort it quickly, and my head begins to swim. I can barely find my way back to the banquette in the darkness, and my heart starts pounding. My fingers tingle, and my eyes are dry. I can't blink. Something is closing in on me, and all I want to do is flee. The inside of my head feels as if it's shrinking against itself, shutting down.

I find Diane and collapse into the banquette next to her.

"Let's go," I manage to say.

"What's the matter?"

Her eyes are glassy.

"I don't know . . . I don't feel so good . . ." I moan.

She dips the corner of a cloth napkin into a glass of ice water and dabs at my forehead. A cold trickle runs down between my eyebrows. I try to take a deep breath, but I can't get in enough air. Nothing's working. Not the champagne, certainly not the coke. I want to take it back, redo the last five minutes, but I can't. Something bad is happening inside me, and I'm terrified.

Diane must have paid our bill, though I have no memory of this, and I hold on to her arm as we stand up. I feel light-headed and for a moment I can't move; I just weave back and forth, watching the prisms of light thrown around the room by the mirrored ball. All this motion is making me nauseated, and I'm afraid I'm going to be sick all over the three men next to us. I feel a hand on my back, and when I turn around the sax player is behind me, and he's offering me a scrap of paper with something scribbled on it. Up close, he's not nearly as cute.

"Later," he says, "we'll be at this after-hours club, if you ladies care to join us."

The street is deserted. This isn't a busy part of town in the first place, but at this time of night not a soul is out walking. The cool air hits my face as soon as we part the club's velvet curtains, and for an instant I feel better. I take a few steps forward, my legs rubbery as a toddler's. My high heel catches a crack in the sidewalk, and I fall into the gutter, tearing a hole through my stockings, skinning my knee.

"Jesus, Dani—how much did you do?" Diane asks.

"Too much," I mutter.

I can't get up. I am sitting on the curb with my head between my knees, and I'm afraid I'm going to vomit all over the street.

Diane stands behind me, reaches under my armpits and yanks me to my feet. She holds on to me, and I feel her breasts pressed into my back. She turns me around and scrutinizes me. She seems to have sobered up fast. Then again, I was the one making all the trips to the ladies' room.

"Let's get you something to eat," she says firmly.

We walk to the corner of Sixth Avenue, where there's an all-night joint called Lox Around the Clock. At this hour it's not exactly hopping. It's a brightly lit room with neon accents, linoleum tables, and the kind of chairs I remember from school lunchrooms. A couple of cops sit in a corner table, hunched over plates of bagels and eggs.

"I don't want to do this—I want to go home," I groan.

"You need to eat," says Diane, who has suddenly turned into Florence Nightingale.

A waitress appears at our table.

"She'll have a bagel and cream cheese, and we'll have two cups of coffee," Diane says in her best Sarah Lawrence Girl voice. I know she's trying not to slur her words.

I'm seeing sparks of light in the corners of my eyes. The world has gone kaleidoscopic. I try to close my eyes, but it gets worse. I blink them open and look across the restaurant at a middle-aged man in a suit and fedora sitting alone at the counter. I think he's my father, and I gasp.

Diane reaches across the table and grabs my hand. "Poor baby," she murmurs. "You'll be okay—I promise."

But I know enough to know that there's no reason to believe her. I have not been okay in so long I have no memory of what it might feel like. I look over at the counter again, and the middle-aged man is gone.

The waitress dumps two cups of coffee on the table so hard they slosh all over the paper place mats.

"Was that necessary?" Diane calls after her.

The waitress hates us—and why wouldn't she? We are eminently hateable. Two drunk spoiled babies. I lean my head into my forearms and begin to cry, even as I tell myself to stop, just stop, because I know if I start there will be no end to it.

My whole body shakes with sobs. In the neon light of Lox Around the Clock, I understand for the first time that I will never see my father again. Time stretches and bends, and becomes something more infinite than my sodden mind can comprehend. Someday I will be thirty, forty even. I will get married, I will have children, and my father won't see any of it. I can't stand the thought that he died worried about me.

Diane comes over to my side of the table and crouches next to me. She wraps her arms around me, her hair falling into my lap. She smells like champagne and coffee.

"Ssshhh," she murmurs. "Breathe."

But I can't breathe, and I can't seem to get a grip on myself. Trying to stop only makes it worse. No matter how much of a mess I've ever been in, there has always been a small piece of me I've held in reserve. That piece is gone.

"Honey, what do you want to do?" Diane asks. She is rubbing circles against my back. I think she's probably scared, too.

The other diners are glancing in our direction, and I overhear someone ask if I'm okay.

"I—"

I can't even get a sentence out.

"Do you want to go home?"

I nod. My whole face is wet, and my insides feel hollowed out. I think about my mother, who is probably asleep in her hospital bed, and that provokes a whole new torrent of tears. I have to live, and, at this moment, that seems like the worst possible thing. What would Lenny do if he was here right now? He would scoop me into his arms and carry me to the car waiting by the curb. He would give me a Valium—which, on top of everything else in my system, would probably land me in the hospital.

"Okay, let's go." Diane leads me out of the diner, carrying my bagel in a brown paper bag. I hold on to her arm like an old woman, and I keep my head down, tears dripping straight to the floor.

She hails a cab on Sixth Avenue and gets in with me.

"I'm taking you home," she says.

It is three in the morning. We don't hit a single light as we ride all the way uptown to my apartment building. Central Park West is dark. Families are asleep, tucked in. This makes me sob harder. My doorman opens the door for us, looking as if he just woke up. Diane takes me upstairs, fishes for keys in my handbag, and fiddles with the two locks on my door. I lean against the wall, still weeping. I think I will weep for the rest of my life.

Once inside, Diane switches on a lamp, then rummages through my dresser drawers until she finds a pair of flannel pajamas.

"Here we go," she says.

She sits me on the bed, kneels in front of me, and removes my shoes. Then she peels off my ripped stockings, skirt, sweater, and bra. She helps me into the pajamas, all the while making little soothing noises, as a mother might to an infant. When she's finished undressing me, she pulls down the bedcovers, plumps the pillows, and tucks me into bed.

I'm never drinking again, I think to myself as Diane sits next to me and strokes my hand. *No drinking, no drugs.* It seems the choice is getting pretty stark. I can't keep going this way. I feel as if my father might be up there somewhere, a pair of eyes in the sky, watching all this. I feel his steady gaze on me.

Diane wipes my cheeks with the sheet.

"Try to sleep," she whispers, but I'm afraid to sleep. If I allow myself to drift off, I might never come back. So I keep my eyes open. I stare at the ceiling, counting backward, reciting the Hebrew alphabet in my head, remembering the faces of my grade-school teachers.

She holds me until the sun comes up.

Lenny will not leave me alone. He calls ten, twelve times a day and leaves messages on my answering machine. *Hello?* His voice fills the shadows of my bedroom. *Fox? Are you there?* Sometimes he doesn't even bother to say a word, but I'm sure it's him. He stays on the line for as long as five minutes, breathing. I know he thinks he can get to me again. As for me, I am trying harder than I ever have before to stay away from him. Sometimes he calls in the middle of the night and wakes me out of a deep sleep, and my hand reaches reflexively for the phone, but I've always snapped out of it just enough to stop myself before it's too late. I'm scared and horribly lonely, but I keep telling myself that anything is better than being with Lenny.

He's in the middle of litigating a famous case, and there's front-page coverage in *The New York Times* and *The Wall Street*

Journal. In these pieces he's referred to as "flamboyant" and "feisty." One reporter likened him to a bulldog, and another referred to his fleet of Ferraris and his trademark raccoon coat. Some people might not want to call attention to this kind of coverage, but Lenny loves it. Just in case I miss any articles and accompanying photographs, Lenny has a messenger drop off clips each evening with my doorman. He's never taken a case he couldn't win—and I guess he thinks he can win me too, if only he's persistent enough.

After all, in the face of the most tangible possible proof that Lenny has been lying to me all these years, I have remained with him. The simple facts about Lenny—who, what, when, where— have always been elusive to me, impossible to grasp as he slipped and slid his way around the truth with all the ease of a snake in a river. *My little girl is dying,* he would say whenever I questioned any discrepancies in his stories, or *I have to go to Houston for a chemotherapy treatment.* A year before my parents' accident, when I couldn't take my own confusion anymore (Was Lenny lying to me? Was I going crazy?), I decided to hire a detective to get to the bottom of it.

When I think back to my younger self riffling through the New York City Yellow Pages in search of a private investigator, I feel that I'm watching a movie about someone else, a girl so clueless that she really didn't know that her desire to hire a detective was all the answer she needed. I chose a detective agency based on nothing more than its good address, in the East Sixties—a neighborhood filled with private schools and shrinks. Most other agencies listed were in the Times Square area or in the Bronx.

It was a cool, crystal-clear spring day when I rang the ground-floor buzzer of a brownstone. A burly, middle-aged man in a sports coat and polyester pants opened the door and ushered me inside. He had a thick head of sandy hair and fleshy pockets under his eyes. He looked exactly like my idea of a detective. I even noticed a trench coat hanging on a standing rack by the door. He pointed me to a small office furnished with a big library desk and two wooden chairs. The desk was covered with papers, half-opened manila folders, the edges of photographs peeking out. An old-fashioned phone, the kind with the dial instead of push buttons, sat on a pile of magazines.

"Mrs. Shapiro?"

"Ms.," I replied faintly.

He blinked.

"I'm John Feeny," he said. "We spoke on the phone."

For this occasion I had dressed as conservatively as I knew how, as if I were playing a role: pants suit, heels, and the pearls Lenny gave me for my twenty-first birthday. My hair was pulled back into a ponytail, around which I had wrapped a silk scarf.

"What can I do for you?" Feeny asked not unkindly.

"You said on the phone that you sometimes deal with . . . personal business," I said. "I have a sort of weird situation."

He leaned back and rested his head in his hands, smiling at me as if to assure me that no situation could possibly be too weird.

"*Ms.* Shapiro," he said. "I was a detective on the New York City police force for twenty-five years before I opened my own shop. And I've been at this here thing"—he waved his hand around the room—"for a decade. So whatever it is, I'm sure I can deal with it."

I suddenly became afraid. Lenny's face floated before me, his eyes bulging with rage, and I remembered a story he liked to tell about how he picked up a broken bottle from the street and slashed a would-be mugger's face. Lenny was sort of a public person. More public all the time. I wondered if I was about to get myself into a whole lot of trouble. But still, I didn't get up, thank Feeny for his time, and head for the door. I stumbled forward like someone trying to escape from a fire. I wasn't sure if I was headed in the right direction, but I knew to keep moving.

"This isn't what you think," I said. "I'm in a relationship with a married man. And I want you to find out if my boyfriend is cheating on me with his wife."

At this, Feeny's eyebrows shot up.

"Come again?"

"He claims his wife is in a mental hospital," I continued. "He told me he hasn't been with her in years."

"And you think he might be lying," said Feeny. Did I see the suspicion of hysteria building behind his eyes, or is my memory supplying it now, because I simply cannot imagine a middle-aged man listening to an earnest, overdressed twenty-two-year-old girl tell him that she thinks her married boyfriend might still be sleeping with his wife?

"Yes," I said.

"What's your boyfriend's name?"

"I'm a bit nervous telling you that."

"Ms. Shapiro, if you don't tell me his name I can't possibly help you. What, is he some kind of famous guy?"

"Well, sort of," I said.

We stared at each other for a moment.

"His name is Leonard Klein, he's—"

"I know who he is," Feeny responded dryly. "The lawyer guy."

I nodded, then sat there, my hands folded in my lap.

"So what do you want? You want him followed?"

"I don't know," I faltered.

"You want pictures? Video? Tape? You want his phone bugged?"

I actually began to get excited. After years of trying to figure Lenny out myself, here, finally, was someone who was going to really do it.

"Everything," I answered giddily. "I want everything."

"It's going to cost you."

"How much?"

Feeny pulled a pocket calculator out from somewhere beneath the mess on his desk.

"Where does Klein live?"

"Upper Westchester."

"So there'll be some travel. You want us to stake out his house?"

"Yes."

"Okay." He peered at the calculator. "I'll need a retainer. And I'm going to have to put a few guys on this. So why don't we say five grand?"

Inwardly I panicked. That amount was about the total I had saved of my own money. Everything else was what Lenny had given me: jewelry, a car, clothes, even cash from time to time. He had gotten me a credit card by lying to the bank and telling them I worked for his law firm. I had steadily been making less and less over the last couple of years, and my parents figured Lenny was supporting me.

"Fine," I said quickly.

I had no sense of whether it was a fair price for what Feeny was going to do for me, or even how I'd survive once I paid him. But I needed to know the truth about Lenny as if my life depended on it.

"Where's Klein now?" Feeny asked as he took the check.

"In Europe," I answered, "on a business trip. He's due back tomorrow on the Concorde from Paris."

"Fine," he said. "We'll start there."

Two days later, Feeny called to tell me that the passenger list for the Air France Concorde—no easy score, he assured me—showed a Mr. and Mrs. Leonard Klein traveling together. And the guy he sent to Kennedy Airport had spotted Lenny and his wife at the baggage claim. The photos were being developed.

I truly hadn't imagined Lenny might be in Europe with his wife. Before he left on this trip, he had given me a hotel number, and an associate in his law firm had answered the phone at the Ritz each time I called. (I later found out that Lenny had made a practice of this: he would fly a Harvard or Columbia Law School graduate across the Atlantic, check him into a hotel room, and give him instructions as to what to say if I called.) Lenny had often told me his European business trips were top-secret: meetings with Margaret Thatcher at 10 Downing Street, fighting the threat of Russian spies.

That afternoon, I called my mother in tears.

"I found out Lenny was in Europe with his wife," I said.

"Oh, darling, I'm so sorry. Is there anything I can do?"

"I don't think so."

A pause.

"Do you want me to call his wife?"

My mother and Mrs. Klein had met each other at a few functions for Sarah Lawrence parents back when none of this could have struck anyone as a remote possibility.

"Yes," I said. "Call her."

"I'll do it right now," my mother said.

I sat by the phone and watched the minutes tick by. I pictured Lenny's wife answering the phone with a chirpy hello, and my mother's slow, steady explanation of why she was calling. I had set in motion a chain of events that was now unstoppable. More than twenty minutes passed before my mother called me back.

"Well, I did it," she said.

"You talked to her?"

The world felt unreal, hallucinatory.

"Yes. She called me a liar. She told me that she has a happy marriage to a man who travels a lot. That he's on his way to California now. And I said, 'No, he's on his way to see my daughter.' "

My mother sounded proud of herself, immersed in the drama of the moment.

"How did she seem?" I asked.

"What do you mean?"

"Lenny's wife—was she angry?"

"No," my mother answered slowly, "she just didn't believe me, Dani."

I spent the rest of that day in a state of awful excitement. *Something* was going to happen. And when Lenny showed up that evening at the apartment we were still sharing in the West Village, I was ready for him.

"How was Paris?" I asked as he put his bags down and gave me a hug.

"Exhausting," he said. "Nonstop meetings."

"Really."

He looked at me oddly, but we didn't have time to get into it. The phone rang. My mother had given Mrs. Klein the number at the apartment and suggested she find out for herself what her husband was up to.

Lenny picked up the phone on the kitchen wall.

"Hello?"

I watched him as his face went white, and for the first and only time in the years I knew him, he looked truly surprised. He didn't say a word. He just listened for a few minutes, then hung up the phone.

"That was my wife," he said.

I was silent.

"How did she get this number?"

I shrugged.

He looked around himself like a suddenly caged animal.

"I have to go."

"I'd imagine," I said faintly. My anger was giving me the fuel that I needed to stay strong, at least for the moment.

As Lenny slammed out of the apartment, I was certain I would never see him again. I knew the truth now. It was staring me in the face,.in the concrete form of flight lists and photos. And besides, the whistle was blown. What could he possibly tell his wife? Knowing Lenny, he'd come up with something. No, this was it, I told myself. Absolutely, positively the end.

Of course, it wasn't the end. And for the next year that we were together—three days here, four days there—I idly wondered what it would take to get me to finally leave Lenny. I wondered about this over bottles of chilled white wine, heavy glasses half filled with scotch. The more I thought about it, the more I drank. And the more I drank, the less I cared.

I was wondering about this at the Golden Door when the phone rang, and in a split second my whole life changed forever.

Dani? Pick up the phone, goddammit.

It is a year later, my father is dead.

No matter how many times Lenny calls, I don't answer the phone.

"Notice anything different?"

My mother beams at me when I walk into her room. She is sitting up in bed, the sleeves of her pajama top rolled up around her biceps, lifting three-pound weights.

"You're lifting weights," I say dumbly.

"Not that, silly."

I look around at the papers on the windowsill, the plant in

the corner, the bulletin board that is entirely covered with photos of my father and me. In one of the snapshots, taken right before I started college, I'm wearing a pink sundress and I'm grinning as my father pulls me close for one of his bear hugs. It's hard to believe that I'm only six years older than the girl in that picture.

"Look at me!" my mother says.

It takes me a minute to register the fact that, after two months, the cast has come off her right leg. I was not prepared for this, and when I first see it, my smile freezes. The skin on her thigh and shin is shiny pink like sausage casing, and her muscles have atrophied. Her right leg is half the size of her left. It doesn't even look human.

I bend down to give her a kiss, then bury my head in her shoulder so she won't see my fear.

"Mazel tov," I murmur. We both know that this is only the beginning. My mother's recovery is going to be measured in degrees: bed, wheelchair, walker, cane. She has not been out of bed in two months.

"I have to finish my exercises. Five, six, seven, eight . . ." she breathes, arms out to her side in a modified butterfly curl.

The hospital can't do anything more for my mother, but she's in no shape to go home. And besides, where is home? It is a question we haven't even begun to explore. Though we haven't said it out loud, we both know she isn't going back to Tewksbury—not now, not ever. A house in the middle of the country with two flights of stairs is not in my mother's future. I have sent away for information about facilities up and down the East Coast, and together we have flipped through glossy brochures depicting manicured lawns and communal television rooms, whirlpools, and state-of-the-art physical therapy facilities. There are no pictures of the residents.

I find it strangely hard to say good-bye to Overlook Hospital.

On my mother's last day, I wander through the lobby and into the empty auditorium, where my father's funeral was held. Slats of dusty sunlight crisscross the darkened room as if from a movie projector. I walk up to the podium and stand in the spot where my father's casket was. I remember the rabbi's thick glasses and long white beard, and the sound of his voice reciting the Twenty-third Psalm. Though I have heard the words only once in my life, they pierce my consciousness like the first few notes of a long-forgotten childhood lullaby. And even though it's not my father's *yahrzeit*, even though there's no religious reason to do this—and in fact it may be sacrilege—I find myself reciting the Mourner's Kaddish.

My father seems a part of this hospital. It was the last place I saw him alive. I'm afraid that when I leave here for good, I will lose him forever. Like the Hebrew words of the Kaddish, what happened in this place will eventually seem both familiar and foreign to me. I will try to hold on to these two months at Overlook Hospital almost as hard as I will try to let them go.

When I drive back to the city on that last night, it's as if my car has a mind of its own. Instead of heading straight for the Lincoln Tunnel, I get off the Turnpike at exit 14 and make my way through the streets of Hillside and pull up outside my childhood home. *Don't do this, don't do this,* I mutter, even as I flick on the blinkers and get out of the car. It's been less than a year since my parents sold this house and moved to Tewksbury. This is the only home I've ever known. Now, as I walk around a row of hedges and peer into the backyard, I see a child's swing set and a station wagon in the driveway where my father's car should have been. The new owners have planted rosebushes along the side of the pool, which is covered by a dark green tarp. A fresh layer of white paint coats the shutters and front columns, making them look too bright against the red brick.

As I stand in front of 885 Revere Drive, time bends and I am a child again. I can see my mother in the backyard doing her needlepoint, my father hanging in traction in the den. There is a pot of goulash simmering on the stove, a housekeeper in a white uniform ironing sheets in the laundry room. I hear the rustling of the elm against the kitchen window, the creak of the bird feeder hanging from its lowest branch, the *thwack* of the mail falling through the slot in the front door. It is as quiet as a museum inside the house of my childhood. I am alone in my room, writing a story about a girl who is already grown-up. I imagine the bright future she has: she is a glamorous doctor, perhaps, or a famous actress. When I finish the story, I shred it into a dozen pieces so my mother won't be able to read it.

The sprinkler system sputters on, and a fine mist blankets the front lawn. I hug myself, shivering in the mild spring breeze as I stand in front of the house in Hillside. It's all gone forever, I think to myself. My father's life—so much of it spent in pain inside these brick walls—is over. My mother is a crippled, angry widow.

The weather has turned warm the morning my mother is transported by ambulance to the Burke Center for Rehabilitation in White Plains. They're moving her at the crack of dawn, so she tells me not to bother driving out to Overlook. When I wake up and look out my bedroom window at the clear, blue sky, I imagine my mother being wheeled on a stretcher through the sliding glass doors of the hospital, and what the air must feel like against her face.

For the first time in a long while, I have the morning to myself. I have nowhere to be and nothing to do—and I'm not sleeping off a hangover. I haven't exactly stopped drinking since that night at Lox Around the Clock, but a subtle shift has taken place inside of me. I *want* to stop drinking—something I've never wanted before. I just don't know how to do it. Every night I go to sleep sober is a small victory.

My phone rings while I'm making coffee. I don't answer it. I never answer my phone. There's still a possibility that it will be Lenny. I know he calls and hangs up all the time, and every once in a while he leaves an angry message. *Goddammit, Dani, I would appreciate the common courtesy of a return call.* So I keep the volume down. I've got a Marlboro in my mouth, a mug of coffee in my hand, and a muffin heating up in the toaster oven before I even remember to check and see who called.

I press the playback button and raise the volume.

"Dani, it's Sheldon." He sounds pissed. Another good reason not to answer the phone. "Listen, kiddo, you didn't even show up at Grey Advertising yesterday for that Special K spot. What the hell is going on with you? Call me."

Sheldon's message drives home the point I've been thinking about for a while now: I can't do it for one more minute. Ever since my parents' accident, the thought of putting on a bathing suit and parading in front of a camera with twenty other girls who all look as if they could be my first cousins seems like the stupidest thing in the world.

I pick up the phone and dial a number in Westchester I still know by heart.

"Sarah Lawrence College," a receptionist answers.

"Dean Kaplan's office, please," I say.

Sarah Lawrence is a small enough school for Dean Kaplan to remember who I am when she gets on the phone.

"Dani, how are you?" she asks warmly.

"Fine," I say.

My voice feels as if it's coming from my toes.

"How's the acting going?"

I close my eyes tight, and my father's face appears. Not the green, haggard postcrash face, but the creased grin I remember from my childhood. *Go ahead,* I can almost hear him whisper. *Take a leap.*

"I'd like to come back," I say. "I know it's been a while, but a lot has happened."

For years to come, when I think of this moment, I will sense my father's hand in this. Until now, going back to college at twenty-four seemed impossible. But I've begun to realize that maybe I can turn back the clock. It's too late for my father, it may be too late for my mother—but I can start over again.

"Well, we'd love to have you," Dean Kaplan says. "Are you thinking of starting this September?"

"Yes."

As she takes down my address and promises to send me registration materials right away, I realize that this is the first thing I've done in years that has made sense.

My mother is sitting in a wheelchair near the parking lot of the Burke Center for Rehabilitation. I'm still not used to seeing her upright. She has a blanket over her lap, and her head is tilted back, angled toward the late afternoon sun. I approach her, casting a shadow over her face. Her eyes blink open, and I see they are filled with tears.

"Where have you been?" she wails.

"I—"

"You're almost two hours late!"

Her mouth is twisted, her face contorted like a child's. When I was a little girl, she was always late picking me up from school. The other kids would leave in their carpools or buses, and I would be left waiting, alone, near the parking lot of a low brick building, scanning the distance for her chocolate-brown boat of an Eldorado to glide into the school's driveway.

"I'm sorry, first I got caught up with some stuff at home, and then I got lost—"

"I'm *helpless* like this," she sputters. "Can't you see?"

How can I not see? Her legs are immobile beneath her blanket, and her feet are propped onto the metal footrests of the wheelchair. Despite all the weight-lifting, her upper body is thin and frail; she probably couldn't even wheel herself up the gentle slope of the sidewalk if she had to.

"Take me inside," she says, imperious as a toddler.

I release the brake on her wheelchair, grip the plastic handles, and begin to push her toward the entrance. It isn't as easy as it looks—the thin wheels sink into every concrete crack—and I feel my mother's weight in front of me as I lean forward, afraid we will hit a bump and she will go flying onto the gravel.

"I want to go to my room," she instructs once we're through the sliding glass doors. She points down a long corridor. Everything at Burke is designed like one big wheelchair-access ramp. The doors are wide, and there are no stairs anywhere. As I wheel her slowly past administrative offices, then turquoise-tiled physical therapy rooms filled with enormous tubs, I keep having the strange sense that something is missing. This place feels eerier than Overlook ever did. After all, Overlook is a hospital—and good things sometimes happen in hospitals. Babies are born and go home. Here, there is no noise, no urgency, no sense of life-or-death. Anyone who has checked into Burke is here because he or she has already survived something unsurvivable. As I push my mother toward her room, we pass fragments of human beings: a quadriplegic boy, a girl whose waist ends in a stump.

"I'm in here," she says, and I turn her wheelchair into what looks like a generic hotel room except for the rails on the bed.

She has already attempted to make herself at home, wedging pictures of my father and me into the corners of the mirror. Three small, tattered books are stacked on her nightstand—my father's prayer books from when he was in the army.

"I'm tired. I want to get in bed," my mother says flatly.

I move the wheelchair right to the side of the bed, park it, then come around front and lift her like a rag doll. She is almost too heavy for me; she can't help herself at all. I get her onto the bed, then lift her legs up and swivel her around until she's lying down. I'm panting by the time I'm through, and she is still looking at me with an expression of ineffable sadness. Anything but this, I think. Let her rage against me, let her scream and yell. But this exquisite blame is more than I can bear.

"Mom, I'm really sorry."

She shakes her head.

"It won't happen again," I say.

I think about telling her about my phone call with the dean at Sarah Lawrence. I know, on some level, that nothing would make her happier than my going back to school. Or would it? I sit at the side of her bed, stroking her hand. The late afternoon light filters through the drapes, casting on everything in her room a yolklike yellow. I open my mouth but the words just don't come out. I can't bring myself to say it: *I'm going back to college, Mom.*

It's bad enough that I will rise from this bed at some point and walk out of here on my own two feet. I will stride down the hall, get into my car, and drive away. But what about my mother? Where will she go, once they have taught her to walk again—if they even can?

I push her hair off her face, and stroke her cheek. Her skin is surprisingly soft.

"I'm going back to college, Mom," I say softly. "I spoke to Dean Kaplan today."

I feel like the sole survivor, emerging dazed from a wreck.

Why should I get to walk away? What have I done to deserve a second chance?

"Oh, Dani," she cries, then pulls me toward her so I can't see her face. "Oh, honey, that's the best news in the world."

The phone rings and I pick it up, for once.

"Hello?" I say gingerly.

There is a beat of silence on the other end, and in that split second my heart sinks.

"Well, well, well. If it isn't the Fox."

"Hi, Lenny," I say faintly.

"You never take my calls," he says, sounding wounded.

"I—"

Why do I need to explain myself to Lenny Klein? I look around at all four walls of my bedroom, as if to remind myself that I'm home, in my own apartment. My door is double-locked, and the doormen downstairs are under strict instructions not to admit Mr. Klein. Even though I know he isn't here, it feels as if he's lurking somewhere in the room.

"Don't you love me anymore?" he asks, his tone implying that this is a rhetorical question, that of course I still love him and I always will. Lenny imagines he has marked me for life, and he probably has, though not in the way he thinks.

I am silent, my mind racing. All of a sudden I can't catch my breath.

Hang up, I think to myself. *Hang up the phone.*

"What do you want?" I ask, sounding smaller than I wish I did.

"I want to see you."

"No, Lenny."

"Just for an hour."

"No, Lenny."

"I'll meet you at the Oak Room bar at six," he says.

I'm in the process of saying *No, Lenny* one more time when I realize the line has gone dead.

He is sitting at a corner table, and I see him as soon as I walk into the bar. He grins at me, then tilts his head back and pops a few nuts into his mouth as I make my way to his table. I realize, horrified, that I still care what Lenny thinks of me—mostly of how I look. Does he like what I'm wearing? Am I thin enough? Does he still think I'm the most beautiful girl in the world? He half rises in his seat, then plops back down when he realizes I have no intention of kissing him hello.

"No kiss?"

I shake my head and fumble through my bag for a cigarette, mostly to give myself something to do. I can't bring myself to look at Lenny, who I know is staring at me with a smirk on his face. My cheeks burn.

A waiter materializes just as I'm lighting my cigarette.

"The lady will have a Johnnie Walker Black—"

"Actually," I interrupt, blowing out smoke, "I'll have a Diet Pepsi."

Lenny looks at me as if I've just ordered a urine sample.

"On the wagon?"

I don't answer him. I'm counting the days between drinks; I haven't been able to go longer than a week before I break down and gulp wine as if it's water.

"I just don't feel like scotch." I shrug. "I've been a little under the weather."

Lenny is still smirking. His eyes travel down my throat, skimming past the open neck of my blouse until he is staring at my breasts. I used to like it when he looked at me like this. It made me feel womanly, sexy. I imagined that Lenny's hunger for me gave me power over him.

"How's your mother?" he asks.

"Better," I respond. This is what I always say when people ask after my mother these days. Better, of course, is a relative term. My mother can now move her right knee three degrees. But Lenny doesn't really give a shit about that.

He reaches across the table and grabs my hand before I have a chance to pull it away.

"I miss you, Fox," he says, his big brown eyes welling up. "Can't we spend a little time together?"

"We're together now," I say, trying to remove my hand.

"Not like this," Lenny gestures around the dark paneled room, which is filling up with the after-work crowd. "Let me take you away. Just two nights."

"I don't think so," I say.

"Please."

I stare wordlessly at the ashtray, my lipstick-smeared cigarette butt. I don't say yes, but I can't quite bring myself to say no.

"One night, then," he says, as if it's settled. "Someplace special. Give me a minute, I'll make a phone call."

He gets up from the table and slides past me, brushing the back of my neck with his hand. I watch him move through the bar and out the door. I know the nearest pay phones are downstairs, that I could, in fact, bolt out of here right now and he wouldn't catch me leaving.

"Would you like another soda?" the waiter asks. I have drained my Diet Pepsi and am sucking on the ice cubes.

"I'll have a Johnnie Walker Black on the rocks," I say.

It is midnight and we are driving north on the Saw Mill River Parkway. I've insisted we take my car. I think I'll have more control over the situation that way. Lenny is behind the wheel, driving with one hand. His other hand has crept up my leg and into my panties, where he is absentmindedly playing with me. My skirt is bunched up, my ass cold against the vinyl seat of my father's Subaru.

"So what have you been doing with yourself?" he asks.

He removes his hand from me and downshifts around a curve.

"What do you mean?"

"These last few months. How have you been spending your time?" Lenny asks.

Is he kidding?

"Oh, I took a little trip to St. Tropez—it's so lovely this time of year," I say.

He looks at me quickly out of the corner of his eye.

"You fucking with me, Fox?"

"No, Lenny, what makes you say that?"

"I was just asking—"

"I know what you were asking," I say. "And I find it incredible."

"And why is that?"

"Lenny, I've been taking care of my mother every day for the past three months. That's what I do with my time."

He falls silent, both hands now on the steering wheel.

"I just meant, what are you doing for fun," he says glumly. "That's all."

I look at his profile as he negotiates the sharp curves of the Saw Mill. He has a soft, jowly face and a thick neck. His hands are stubby, carefully groomed, each nail filed into a perfect pink half-moon. He's still wearing his suit, and his stomach bulges softly through his suspenders.

"I stopped having fun," I say.

"That's too bad," he replies.

It seems the three scotches I downed before we left the city have done nothing to dull this edgy feeling that I'm doing something stupid and wrong. I have no business being in this car with Lenny Klein, heading for God-knows-where.

"And you?" I ask. "What have you been doing for . . . fun?"

"I don't think you really want to know," Lenny says.

This, he knows, will get me going.

"Oh, but I do," I say. "I do want to know."

"Well, I spent some time in LA," he says.

"And?"

"Well, I spent some time with a particular young actress in LA," he says slowly.

He mentions a name, and it's one I—and the rest of America—can put a face to: a beautiful, painfully thin blonde who plays an upper-crust vixen on one of the nighttime melodramas.

"Are you sleeping with her?" I ask.

I don't stop to think about why on earth she would sleep with him. Years from now, while running in Central Park, I will pass this young actress, also running. My pulse will race even faster, and I will fight the urge to stop her and ask. But what to ask? *Excuse me, but this crazy ex-boyfriend of mine once told me he had an affair with you. Does the name Lenny Klein ring a bell?*

"Here we are," says Lenny as he pulls the car into the driveway of a small inn, which looks closed for the night. He parks at an angle next to a tree, taking up two spaces. He starts to get out, but pauses when he sees that I'm not moving. I stare straight ahead. With the lights on in the car, I can see my own reflection in the windshield, an indistinct shape with no beginning or end.

"Are we waiting for something?" he asks somewhat impatiently. I wonder how Lenny handles his eight children.

"I asked you a question," I say.

"What?"

"Did you sleep with her?"

This time my voice has acquired an edge.

"Yes, I slept with her"—he shoots back at me. "What did you think? That I was living like a monk?"

I close my eyes, images of Lenny in bed with the Hollywood vixen-of-the-moment flashing like cue cards. Once, this would have turned me on. I liked the idea of Lenny sleeping with beautiful women—it made me feel like less of an idiot for being with him. Lenny has narrated his whole sexual history for me, knowing that with each movie star, singer, dancer he claimed to have bedded, his stock rose higher and higher. Whether these stories were anything more than a forty-six-year-old man's wet dreams has never been something I've considered.

I take a deep breath and get out of the car. All I'm carrying is a small purse containing a twenty-dollar bill, my American Express card, and a lipstick. The temperature has dropped since we left the city. I hug my thin sweater around myself. Suddenly I'm exhausted, and all I want to do is sleep.

"The key is supposed to be under the mat," Lenny says as he walks ahead of me, then bends forward and rummages around in the dark. He finds it, fiddles with the lock, and lets us in. There is a small lamp lit in the foyer, and an envelope with his name on it propped next to a vase of flowers on the reception desk. He motions me up a narrow, carpeted staircase. I wonder if Lenny has been here before. He certainly seems to know his way around.

As soon as Lenny closes the door behind us, he pushes my skirt up around my waist and yanks my panties down with his other hand.

"Oh, Fox," he moans, nudging me backward, onto the bed. "I've missed you."

He unbuttons my sweater, his fingers fumbling. He is lying on top of me now, his fly somehow unzipped, and he's pressing himself into me. I can barely breathe. His stubble is scratching my face, his breath smells of something garlicky, and I turn my head to the side to stop from gagging.

"What are you thinking?" he whispers in my ear. I know what he wants. He wants me to talk dirty to him, to remind him that he's in bed with a girl half his age, but I just can't do it.

"Nothing," I whisper back.

There are weathered beams running across the ceiling, and I count them as Lenny moves on top of me, crushing me. I want to scream at him to stop, to push him off, but instead I just lie there, flopping around like a rag doll.

Finally he lets out a groan and collapses. For a moment I wonder if he's dropped dead of a heart attack. But then he lifts himself up and rolls over. I try to breathe.

He leans on one elbow and looks down at me. Pieces of hair are plastered to his forehead.

"What's the matter with you?" he asks.

"What do you mean?"

"You're like a dead fish," he says.

I move away from him.

"Thanks a lot."

"It's like you aren't even here," he says.

"Why don't you just call your little actress friend," I say. "I'm sure she can do better."

"I may just do that," says Lenny, heaving himself to a sitting position.

"Go right ahead," I mutter, and curl into a little ball, my back to him.

I try to sleep, but rage keeps me awake. I feel blood racing through my veins, pounding in my ears. The loneliness I've felt since my father's death is nothing compared to this. Lenny leaves the room at some point, then returns, his breath smelling of whiskey. He reaches for me, a hand on my hip, and I shake him off violently. I can't bear his touch. My side of the bed faces a window, and through the sheer white curtains I see the sky littered with stars, an almost full moon. Branches rustle against the side of the inn, a sound I imagine would be comforting and romantic if I were wrapped in the arms of someone I loved. Lenny has passed out on top of the covers.

As I watch the shadows play against the wall, I think about the past three months. Try as I might to empty my mind, I find myself reliving the moment I received the phone call, the first time I saw my mother's face after the crash, the image of the plain pine box with my father inside it being lowered into the ground.

When the first pale orange streaks of dawn finally appear across the sky, I climb out of bed and try to make myself presentable—no small feat, considering I slept in most of my clothes. I smooth my skirt and button my sweater, splash my face with cold water. When I walk over to the bed, I see that Lenny's eyes are open.

"Let's go," I say. "I want to go home."

Without a word he gets up, pulls on his suit pants, lifting his suspenders over his shoulders. He picks up his briefcase and stands by the door, glaring at me. His eyes are dark pits in the early morning light. We make our way quietly down the narrow staircase and out the front door. In the parking lot, he moves toward the driver's side of the Subaru, but I stop him.

"I'm driving, Lenny," I say.

It's seven on a weekday morning, and already there are commuters on the road. I follow signs to Route 684, which will be the most direct way back to the city. I turn on the radio; Lenny turns it off. I smile a tight, sarcastic little smile.

"You have no idea what you're losing," he says after a while. I have cracked the driver's side window, and the air rushing in smells impossibly fresh.

"Oh, I think I know exactly what I'm losing," I say.

With each mile we drive toward the city I feel lighter. Even though I'm dehydrated and hungover, even though I've been up all night tossing and turning in my clothes, I'm more at peace than I've been in months. I have no idea what the future holds for me, but I know what it *doesn't* hold, and right now that seems like a small miracle.

We are speeding down Route 684, approaching exit 4 when Lenny tells me to pull over.

"What do you mean?" I ask, my fingers wrapped tightly around the steering wheel.

"I want to get out," he announces.

"But—"

"Just do it," he snaps.

I pull off the highway next to the exit ramp, my hazards flashing, and as I do so I realize that Lenny doesn't live far from here, that in fact he probably is within a few miles of home. I can see he's worked himself into a total froth. His nostrils flare, and his eyes are bulging more than usual.

"I won't be calling you again, Fox," he says.

He waits a millisecond for me to change my mind.

"That's it, then," he continues.

I look at him calmly.

This is how I will always remember Lenny Klein: he gets out of the car and begins walking up the exit ramp, wearing a navy-blue pin-striped suit and carrying a brown leather brief-case, his thick black hairweave almost ripping off his head in the strong May wind. I imagine his getting to the top of the ramp and—what? Sticking his thumb out? Stopping at a pay phone and calling his wife? Concocting some story about abduction and blackmail?

I carefully pull back onto the highway. I turn on the radio and scan the stations until I find something classical, calming. I can almost hear my father's voice in my head. *Beethoven*, he says, *the third piano concerto*. Tears are rolling down my cheeks, but I'm not sad, not about Lenny, at least.

CHAPTER
EIGHT

The address is scribbled on a scrap of paper, tucked into the bottom of my wallet, wedged beneath dusty pennies, Chinese fortunes, stray Tic Tacs. *Church of St. Paul and St. Andrew, Eighty-sixth Street and West End Avenue.* I stand a few feet away from the church, sweating on this humid August night, pretending I'm waiting for someone on the corner. As if anyone notices. As if anyone cares. My stomach lurches; I'm sick with fear. I start to walk away, then stop.

I have been carrying this address around for over a year. It was given to me by a girl I met in acting class who saw the way I drank each time a group of us went out for a few beers. *Listen,* she said to me once, pressing the scrap of paper into my palm, *this may be none of my business, but maybe you want to check this place out.* I stuffed it into my wallet, and it was weeks before I looked at it: *Alcoholics Anonymous,* she had scribbled, along with a few names of churches and times of meetings. Fuck her, I thought. What does she know? And then I neatly folded the paper and left it in my wallet.

Just before six-thirty, I open the side door where there is a small sign with an arrow pointing down a narrow flight of stairs to the basement. I've been inside a church only once in my life, for the funeral of a friend who died in a drunk-driving accident in high school. The crucifix above the door makes me even more nervous. I have no right to be here. This is a place for goyish alcoholics, not a Jewish girl who maybe drinks too much. I wonder what my father would think.

While I've been standing outside trying to make up my mind, dozens of people have walked in, but I can't imagine they're going to an AA meeting. They look healthy—happy, even. I'm expecting bums in raincoats, old red-nosed men. AA hasn't quite achieved the celebrity status that it will over the next few years. It isn't cool to be an alcoholic. Actors aren't joking about it on *The Tonight Show*.

I want to be invisible as I sidle into the brightly lit church basement. I'm not even sure I'm in the right room; this looks more like a college reunion than a meeting of Alcoholics Anonymous. *Alcoholic*: it's a word I've never even thought, much less spoken out loud. My heart is pounding. I swallow hard over the lump in my throat as I scan the crowded room for a seat.

A woman about my age holding a cordless microphone stands in front of the room.

"Hi, I'm Mary and I'm an alcoholic," she says.

"Hi, Mary!" the whole room responds in a resonant boom.

I nearly jump out of my seat, and the man next to me notices and leans over.

"This your first time?" he asks.

I shake my head no and hunch down in my folding chair. I don't want anyone to talk to me. I want to bolt, but I'm afraid if I do they'll come after me with a giant net they reserve for fleeing first-timers. They'll drag me back in and the whole room will point at me and laugh. I sink deeper until I can barely see over the head of the woman in front of me.

Mary the alcoholic starts to tell the story of her childhood. As she talks her words move through my body like water, leaving trace elements I will remember years from now. She's from a big Irish Catholic family; six out of seven of her siblings are heavy drinkers. Her father died of alcoholism. I picture a dim Irish pub with a neon four-leaf clover in the window, and Mary's red-haired brothers and sisters knocking back shots of whiskey at the bar. *See, I'm not an alcoholic,* I think to myself. *Jews aren't alcoholics.* I think about my family, and how almost nobody drinks. My father would have a thimbleful of wine on Shabbos, my mother barely touched the stuff. I don't consider my father's drug addiction or the vodka on Uncle Harvey's breath.

There's an older man across the aisle from me and I think I recognize him from a sitcom I used to watch as a kid. He's leaning forward, elbows on knees, intently listening to Mary. Everyone seems focused except for me. I count sixty-seven heads in this room. All the folding chairs are filled, and there are people sitting on the floor. This basement must be used as a nursery school during the day: the walls are covered with cutouts of trees and flowers. Each leaf is inscribed with a child's name and future occupation: Jennifer wants to be a nurse when she grows up so she can help sick people. Zachary is planning to be an astronaut so he can fly on the space shuttle.

I feel alien in this crowd—a terrible feeling, and a familiar one. What am I doing here? Okay, maybe I drink too much, but this is a little extreme. I can't do things in moderation: either I drink until I'm falling-down drunk or I find myself at Alcoholics Anonymous. Isn't there a middle ground? Can't I just impose some self-discipline, cut back a bit? These people look like Moonies to me, with their shining eyes. I feel trapped. I look at my watch. Forty-three minutes and this meeting will be over. My chair squeaks.

I try to listen to Mary. She's saying something about white wine spritzers that makes people laugh. What's so funny about white wine spritzers? A poor excuse for a drink, if you ask me. She talks about the way she used to break into a cold sweat in the middle of the night, and hate herself in the morning. I wake up in cold sweats all the time. Doesn't everybody? Now she's talking about how booze fueled her choices of schools, jobs, men. Lenny's face appears before me. If I hadn't been drunk whenever I was with him, would there have been any possibility I would have stayed with him for nearly four years? I try out this equation in my head: Lenny without alcohol. It doesn't compute.

I'm sweating in the sweltering heat of this basement. The people around me don't seem to be as uncomfortable as I am. I run my fingers through my damp hair. I can't sit still. Now Mary's talking about how she'd notice black-and-blue marks on her body and not know how she got them. Everyone laughs, and I want to scream. I look down at my legs, wishing I hadn't worn shorts tonight. I've been telling myself my bruises have to do with a vitamin C deficiency. Is Mary a plant? Did they somehow know I was coming, and hire an actress to pretend that this is her life, when really it's mine? How does she know what's been going on inside of me?

When she finishes speaking, everyone applauds as if we're at the Philharmonic. Then a guy in a baseball cap takes the microphone and announces that AA has no dues or fees, but it does have expenses, so he's passing a basket around the room. Most people throw in some change or a dollar bill. When it comes to me, I put a crumpled dollar on top of the pile.

"This is a closed meeting of Alcoholics Anonymous," he

says, reading from a card. "Attendance is restricted to those with a desire to stop drinking."

A *desire to stop drinking*. I think of the lists of resolutions I have scribbled in my journal each morning. Cutting down on my drinking has been at the top of the list for as long as I can remember. I can't imagine stopping. But it seems that's precisely what the people in this room have done.

"Is anyone here for the first time who would like to say hello?" the baseball-capped guy asks.

I feel that everyone is looking at me. I stare at a coffee stain on the floor.

"Okay . . . how about anyone counting days?"

A bunch of hands shoot up.

"Hi, I'm Stan and I'm an alcoholic and today is fifty-three days," says a guy behind me.

Applause.

"I'm Debby, alcoholic, and I have three days back," a long-haired girl announces, smiling bravely.

More applause.

A dozen people introduce themselves and state the number of days they have, ranging from three to ninety.

The meeting goes on for what seems like forever. People raise their hands, and Mary calls on them. The cordless mike is passed around the room like a baton. Words drift through my consciousness like passing clouds, words like *powerless, denial,* and worst of all, *higher power.* It's as if there's this whole other language in here, a blend of psychobabble and spiritual mumbo jumbo. I've spent enough of my life in stained-glass sanctuaries listening to rabbis ponder the meaning of God.

I close my eyes and think about my father. I do that a lot these days—think about him whenever I'm hurting inside. As usual, I can't see his pre-crash face right away. I'm plagued with images of his vacant eyes, stubbly cheeks, wire-rimmed glasses askew on his face. I concentrate, focusing hard until he finally comes to me—until he's so real that I can hear his voice and feel his touch.

So what do you think, Dad? Your daughter's in a church basement listening to people talk about booze. Pretty silly, huh?

I'm not so sure, shnookie. His voice is as clear to me as the voice of the guy speaking into the mike. *I'm not so sure.*

When the meeting ends, everyone applauds, then stands, and the people on either side of me grab my hands. Recoiling will only mark me as new, and I don't want to give myself away. So I hold hands, and, for a moment, become part of this human chain as Mary says she'd like to close the meeting with something called the Serenity Prayer.

"God," the whole room intones, "Grant me the serenity to accept the things I cannot change, the courage to change the things I can, and the wisdom to know the difference."

I try to bolt the second the meeting is over, but my path to the door is like an obstacle course.

"I haven't seen you before. Are you new?" an older woman asks.

"No, I'm visiting from California," I mumble as I keep moving toward the exit sign.

"Hi, I'm Brooke," a pretty girl says. "This your first meeting?"

How do they know? Am I wearing it like a sign across my forehead? People here seem to have my number. I keep my eyes on the scuffed-up linoleum floor as I make my way out.

When I finally push the heavy doors open and escape into the night, I'm so glad to be out of there I barely notice the oppressive heat as I round the corner quickly and begin to walk down West End Avenue.

Never doing that again.

Why not, shnookie?

They're a bunch of loonies.

Maybe those loonies can help you.

I argue with my father in my head all the way home. I hated every minute of that meeting, and yet, now that it's over I feel oddly lighthearted. Something about hearing all those stories. Never in my life have I heard so many people exposing themselves to one another, seemingly without fear. I felt like a voyeur while I was there, but now that I'm out on the street I feel comforted. It seems to me that the stories I heard in the meeting are part of a vast universe of stories. No one has it easy—not really. I think of the pretty girl who tried to be nice to me as I was leaving. Whatever brought her through the doors of Alcoholics Anonymous couldn't have been pretty. I'm not the only one having a hard time.

Maybe I'll just try to stop on my own.

You haven't had much luck with that.

But—

No buts.

My father's voice is firmer than it was in life, somehow more full of conviction. On my way home, I pull the scrap of paper from my wallet to see what else is written there. Tomorrow night there's a meeting at the Church of the Heavenly Rest on the other side of town.

It is eight o'clock on the Tuesday after Labor Day weekend, and I'm in the garage in my building waiting for my car. I'm on my way to Sarah Lawrence College for the first day of classes, and I feel the way I did in seventh grade, when my parents took me out of the yeshiva and put me in prep school. There is a chasm in front of me, and all I can do is take a running leap.

That I'm doing this without the aid or benefit of a few drinks makes me feel doubly exposed. I'm twenty-four years old, and the thought that I'm going back to college after everything that's happened is the most terrifying thing I can imagine.

I'm sweating, my whole body damp under the clothes I carefully chose for today. I close my eyes and a wordless, formless prayer zips through my head, attaching itself to nothing. *Please help* is the gist of it. I don't know whom I'm praying to: God? My father? They both seem out of reach today.

I try to breathe deeply, but the fumes of the garage make me gag. I can see myself in the security monitor: I'm wearing jeans, cowboy boots, a navy-blue cotton sweater. Suddenly I feel like a complete dork. Does anyone wear cowboy boots anymore? It seems I've been out of college for a hundred years. When I tried to figure out what to wear this morning, I felt as if I was dressing for a role: *college-girl type*. By the time I left, clothes were strewn all over the bed: short skirts, sweaters, dressy pants, high heels, designer jackets.

As I pull out of the garage a black Mercedes sedan is pulling in, and I see a woman's tanned arm resting against the open window, a gold watch on her wrist. The watch catches my eye— it's the same as the one Lenny bought me in Monte Carlo a few years back. The woman's face is partially hidden behind enormous tortoiseshell sunglasses, but the clean line of her jaw and the brown sweep of her pageboy are unmistakable. It has been seven months since Lenny and I broke up, four months since the last night we spent together, and today I am looking at the cold, well-kept beauty of his wife.

What the hell is she doing here, in my garage? I drive down Seventy-third Street and double-park outside the Korean deli on the corner of Columbus. I have an awful feeling that Lenny may have moved his family into the city. I fumble for a quarter, then call directory assistance and ask for Leonard Klein. And when the operator reels off a Central Park West address one block from where I live, I hang up the phone, shaking. Am I going to run into Mrs. Klein at the greengrocer? See Lenny taking his kids for Sunday walks?

In the deli, as I wait for my coffee and bagel, I imagine Lenny as a giant boulder, then heave him away from me with all my might. I picture him rolling downhill—fast, faster. So what if he's moved his whole family a block away from me? I deserve nothing less. I get back into my car and make the twenty-five-minute drive to Sarah Lawrence, my mind racing. Did he do this to get to me? Is it possible that the only perfect apartment in all of Manhattan was one from which he could toss a stone and hit my bedroom window?

The sight of the George Washington Bridge stretched across the Hudson River and the majestic buildings of upper Riverside Drive does nothing to lift my spirits. My rage at Lenny turns into a boomerang, which I level back at myself. As I press my foot to the gas and push the Subaru to seventy, I know it's my own fault this is happening. And this great idea of going back to college— it's a waste of time and money. What am I going to do with my life, after all? I should turn the car around and go home. I want to pour myself a drink and get back into bed. All I am is a pretty girl who has used her looks to get into a whole lot of trouble— and that's all I'll ever be. I spent four years with a married man, and God is going to punish me.

It's all I can do to keep my car pointed in the right direction. I drive past the bridge exit, through two tollbooths, and before I know it I am on the campus of Sarah Lawrence College, where once, not a very long time ago, I was young.

I park the car and walk up a narrow path to a Tudor building shaded by an enormous old tree, its back windows overlooking a clay tennis court with a sagging net. This is the dorm where I was living when Lenny Klein first showed up in my life. This is where Jess and I used to stay up late at night and help each other with term papers. I see her slouched on the floor of my room with loose pages of research spread around her, her hair pulled back in a rubber band, the soles of her feet dirty from running barefoot across campus.

I lean against the side of the building and try to calm myself. It's been two weeks since my first AA meeting, two weeks since I've had a drink, and my inner life is like a battlefield, scarred, unrecognizable. I can't seem to stop my feelings, whatever they are, from taking over my body. I feel great sweeps of rage, fear, impatience, guilt, anxiety.

I shake my head hard, trying to get rid of the images flooding through me. Before I open the front door, I imagine myself, four years younger, running through it and into Lenny's waiting car. I walk through the lobby and into the back of the building, where there are faculty offices. There's one class I promised myself I'd try to get into, taught by Ilja Wachs, one of the most popular and respected professors on campus.

What makes me think I have a prayer of getting into his nineteenth-century literature class? My reading for the past four years has consisted of bad scripts and *Women's Wear Daily*. But I have so much to prove to myself—everything, in fact—and this seems like a good place to start: Flaubert, Tolstoy, Eliot, Dickens. His course description made clear that his class will read a book a week.

Even though I think he'll probably laugh me out of his office, even though I can smell the fear on my own body, I knock on his open door and poke my head inside.

"Come in, come in."

The large, white-haired man hunched over his desk doesn't look up as he waves me into his office. He's looking for something on his desk, which is covered with papers, ceramic coffee mugs, an overflowing ashtray. There are bookcases lining the walls and knee-high piles of books all over the floor. The room smells as if he's already smoked a pack of cigarettes this morning.

"Here it is," he mutters, extracting a single sheet of paper from beneath a mountain of pages just like it. Only then does he look up, his blue eyes piercing and kind.

"And who are you?" he asks.

I stutter my name.

"What can I do for you, Dani Shapiro?"

He leans back in his chair, arms clasped behind his head. The buttons on his oxford shirt spread apart, a pale, gray crescent of skin peeking out.

"I'm interested in taking your nineteenth-century literature class," I say.

He smiles, his face crinkling.

"Why?"

Why? I wasn't expecting this. I feel as if I'm on an audition, but I don't have a script. What does he want to hear? I'm so used to telling people what I think they want to hear that I have no idea how to talk about my own thoughts and feelings. I feel my cheeks flush, and I wish I'd just sink through the floor and disappear.

"I—"

I don't know what to say. I trail off, staring out the open window. Two women in shorts and sweaters are swatting a tennis ball back and forth.

Ilja doesn't take his eyes off me, nor does he help me out. But somehow I know I'm in a safe place. By walking through the door of his office, I seem to have put myself in his care. I remember that he was the adviser of my freshman-year roommate.

She would come back from conferences with him, her eyes red-rimmed and puffy, after telling him her problems academic and otherwise.

"Well, what have you read recently?" he asks.

"I've been out of school for a while," I say.

He raises his eyebrows.

"I dropped out in '83," I say, "to work as an actress."

"And did you?"

"Sorry?"

"Did you work as an actress?"

"Yes," I respond, and then begin to reel off my list of credits until it occurs to me that I'm talking to someone who isn't going to be impressed by my SAG card or string of national television commercials. He probably doesn't even own a television.

"What made you decide to come back to Sarah Lawrence?" he asks.

"My father died seven months ago," I answer haltingly.

Ilja leans forward in his chair, suddenly, completely, focused on me. The room is very quiet. All I hear is the rhythmic thwack of a tennis ball, a few chirping birds outside the window.

"Tell me everything," he says.

And so I begin. I tell this man I've never met before the story of my life. I tell him about my dead father, my crippled mother, my relatives who now communicate with one another only through lawyers. I tell him about Lenny Klein, and about how I feel I'll never be young again. He pushes a box of Kleenex across his desk, papers flying to the floor, and he rests his chin in one hand, listening, his white hair flipped to the side like an inverted comma. I feel that he's actually interested in me, that I have all the time in the world. There are places in my story where he barks with laughter, others where shadows cross his face.

"Oh, my dear girl," he says simply, when I've finished.

I'm expecting him to refer me to the school shrink or suggest I rethink my college plan. I bow my head, shredding a piece of tissue balled in my hand.

"My class meets Tuesday and Thursday mornings at nine," he says.

I look up at him.

"It isn't going to be easy for you," he says. "But we'll work together. Can you come see me once a week, for an hour before class?"

"Yes," I answer. What must he see in me? I cannot even begin to ask the question, much less answer it. I stand up, dizzy on my feet, and he clasps my hand between his two warm palms, cupping it like a leaf in the wind.

"Welcome back," he says.

"Three bedrooms, river view," pronounces the real estate broker, who is standing in her Nancy Reagan–red suit in the center of an empty living room, her heel tapping against the parquet floor.

This is the sixteenth apartment she's shown me this week, and she's beginning to get impatient. I'm looking for an apartment for my mother, and I have a list of her specifications: door wide enough for a wheelchair, no stairs in the lobby, a kitchen and bathroom she can use without too much trouble, and enough space for a live-in nurse. My mother is used to a suburban house. She needs a big apartment—and not just any big apartment. She also wants central air-conditioning, a twenty-four-hour doorman, and an Upper West Side location. She wants to live near me.

I pull out my tapemeasure and prowl through the rooms, checking the dimensions of the doorways, the height of the counters.

"It's certainly *bright*," says the broker. She puts on her sunglasses to make her point.

"Mmmpphh," I mumble. I wish she'd just leave me alone for a few minutes. Whether my mother can negotiate her way through this space in a wheelchair is only part of what I'm thinking about. I'm trying to picture her here. In my mind, I fill the rooms with the furniture from Tewksbury, I hang art on the walls. Will this place feel like home? Will she curl up and watch television at night, comforted by the lights of New Jersey across the river?

"So what do you think?" The broker sidles up to me and rests her cool, manicured fingers on my arm. "Will Mom be happy here?"

A flash of anger rips through me, and I want to lash out at somebody—but at whom? I see myself reflected in the broker's sunglasses, distorted, moonfaced. I feel as if I'm disappearing, as if my sole purpose in life is being my mother's daughter. Wherever I go, people ask me how my mother is. *How's Mom?* Concern flickering across their well-meaning faces. *How's she holding up?* They never ask me how I'm doing, how I'm holding up. No one seems to realize that I've barely begun to grieve for my father. Since the accident, taking care of my mother has fueled me. Now, I spend half my weeks in classes and AA meetings and the other half talking to doctors, lawyers, real estate brokers until I'm afraid I'm going to split in two.

I stretch the tapemeasure across the entrance to the master bedroom. If we take the door off its hinges, she'll be able to make the turn from the hallway and get her wheelchair inside. I look around the empty room and try to imagine a bed, a bureau, curtains on the windows overlooking the smokestacks of Hoboken. Will my mother be happy here? Who am I kidding? My mother may be moving into a three-bedroom apartment, but happiness is a luxury she can't afford.

"Fine." I turn to the broker, who is furiously flipping through a large week-at-a-glance ledger. "We'll take it."

The morning the movers arrive, I am standing under the awning of my mother's new building, trying to quiet the little voice in my head that is telling me I've made a terrible mistake. My mother is going to hate her new apartment, and it will be all my fault. I think I'm an awful daughter, uncaring and ungiving. If only I had kept looking at another ten or twenty apartments, I might have found the perfect one.

"Where do you want it, lady?" a guy calls from the back of the truck, pointing to a beige sofa.

"I don't know—let's just get everything inside," I say. I think about the way my friends talk about going home to see their parents. There will be none of that for me. The autumn sunlight is having a strobe effect on my vision, and everything looks jerky, like an old movie.

I watch as the movers cart in the sofa, armchairs, television, coffee table, and bed. It's all rented furniture—every last stick of it. My mother didn't want to deal with the stuff in storage from the house in Tewksbury, so she asked me to rent everything. From the rolled-up rugs the movers are carrying inside to the flatware and china carefully packed in boxes, there isn't a single object going into my mother's apartment from my parents' life before the accident.

Once everything is inside and the movers have been paid, I begin to unpack my mother's new things. I unroll the rugs, find the glasses, dishes, flatware, and put them where I think she'd want them. I check the phone to make sure it's working, along with the gas, and the cable television. My mother is due to arrive by ambulette tomorrow, and as much as possible I want this place to look like home.

By the time I've finished, the sun is setting over the Hudson

River, casting an orange light across the living room, so intense I lower the blinds. I survey the room. Not bad, I think. It looks like an upscale motel room—probably because beige seems to be the rental-furniture color of choice.

I take the elevator downstairs, and walk one block over to an outdoor market on the corner of Broadway. There are dozens of different colored tulips nestled behind crates of oranges, plums, grapefruits, nectarines. I grab a plastic basket and begin filling it up with fruit and flowers. I take it back to my mother's house, locate a vase, and fill it with tulips. Then I dig through some wrapping paper and find a glass bowl for the fruit. I put the vase and bowl on the beige formica coffee table, turn out the lights, and close the door behind me.

It's time to order my father's tombstone. A simple matter, really, since the other graves in the family plot are marked in a uniform way: small stones etched in Hebrew stand at the foot of all the graves, which are covered with ivy.

My mother wants to plant a hedge on top of my father's grave. Not only does she want a hedge but she wants his tombstone to have English on it—words that describe how we felt about him.

"How about 'devoted husband and father'?" she suggests.

"Ugh."

" 'Businessman and philanthropist'?"

I shake my head so hard it rattles.

"Well, what about 'a learned man of Torah'?" she offers. "He would have liked that."

It all seems wrong to me, but I can't quite put my finger on the problem. When my mother finally comes up with a novella-length epitaph—*Truly the son of Joseph in his own time, he taught us wisdom, kindness, and above all, compassion*—I agree, exhausted by the effort of trying to sum up a life in a few words. It will be years before I realize that what my father really would have wanted was a small stone etched in Hebrew, just like the rest of them. The hedge covering his grave will grow higher than the ivy-covered plots surrounding it—making a spectacle of my father in death, the way he was always careful not to be in life.

The unveiling is planned for the last Sunday in October. *Unveiling*: a word I have heard before without considering its meaning. I imagine a scarf—sheer, filmy, white—floating above my father's grave, settling on top of it like a butterfly net.

"So who'll be there?" I ask my mother.

Now that she's living in the city and wheeling herself around, she's begun to plan things herself.

"You and me, Morton and Shirley Sugerman, Roz, a few friends . . ." she trails off vaguely.

The horror begins in my toes and moves slowly up my body, ice-cold and unstoppable.

"Anybody else?" I ask.

"Not that I can think of," my mother answers.

"What about Susie?"

"Oh, yeah. Susie," she says.

My mother and half sister have stopped speaking altogether ever since Susie contested my father's will. Eventually, Susie will lose this battle, but she will have held up my father's estate for many months, costing my mother a lot of money. I should have known that for Susie this is largely an emotional battle; she is still the nine-year-old girl whose father went away, and she wants a piece of him, if only a financial one. And as far as my mother is concerned, she has never felt accepted or loved by my father's family, and now that he's gone there's no reason for her to pretend she even likes them. Sometimes I wish I could lock them all in a room. What would be left of them? Bits of hair and bone?

"And Shirl and Harv?" I ask. Surely my father's sister and brother will be at the unveiling.

"No," she says. "They're not coming."

"Not coming, or not invited?"

"Not invited," she says, looking just past me.

How can you do this? I want to scream at her, but I don't. Fighting with my mother gets me nowhere. She is utterly, absolutely certain that she is, in all matters, victimized and therefore entitled to retribution. The accident has deepened her belief that my father's relatives are out to get her. For years to come, when their names come up, she will mutter, *They almost killed me.*

"Are you sure about this?" I ask. "I really think—"

"Not invited," she repeats.

"But, Mom—"

"Your father wouldn't have wanted them there. Not after what they've done to me."

"You're talking about his sister and brother," I say.

"So what," she says, her eyes dark and defiant. "They didn't care about me, so I don't care about them."

When I pull up in front of my mother's building on the morning of the unveiling, she's parked just outside the front door in her wheelchair, wearing sunglasses and a dark tailored suit. She waves when she sees me, smiling as if we're going to lunch, not the cemetery.

Without a word, I release the break on her wheelchair, then push her to the corner, where the curb flattens for handicapped access. I help her into my car, then collapse the wheelchair and put it in the trunk. As we drive to Bensonhurst, my anger at my mother all but disappears. I'm thinking that this is the first time she'll be seeing my father's grave. The rabbi has told me that the spiritual reason for an unveiling has to do with closure, but for my mother I'm afraid that seeing the grave will mark the beginning of grief—not the end of it.

"Is there anything you want to say?" my mother asks as we whiz through the Brooklyn Battery Tunnel.

"You mean now?" I respond. I've barely spoken since we got in the car.

"At the service. Do you want to say a few words?"

"No," I say quickly, without even thinking. I never considered the possibility that people might speak at this thing. "Why, is anyone else saying anything?"

"Well, I'm planning to," my mother says. I glance at her and notice for the first time that she's holding some single-spaced typed pages.

"Maybe Susie will want to say something," I say casually. Out of the corner of my eye I watch my mother stiffen and feel a momentary rush of satisfaction, immediately followed by guilt.

We drive through the gates, where Roz, Morton, Shirley Sugerman, and a few of my mother's old friends are standing in a cluster. Off to the side, Susie is leaning against the office wall talking to the rabbi. Her hair whips around her face in the warm breeze, her mouth tight and grim. She almost didn't come to the unveiling when she found out my mother excluded Shirl and Harv.

"Hi, Dan-Dan," Roz chirps.

"Dan-o," Morton kisses me on the cheek.

They're acting as if nothing is wrong. Don't they realize that the ties connecting me to my father are being systematically snipped? To them, the conspicuous absence of my father's family seems to be a small, incidental fact. I tell myself I'll go talk about this in an AA meeting later. I'll raise my hand and share about how my mother banned my father's family from his unveiling, and the strangers in the room will nod their heads sadly, letting me know they understand.

I lift the wheelchair from the trunk, snap the safeties into place, then help my mother out of the passenger seat and into the chair. She has begun to take a few steps with a walker, but this is no time to experiment. Gravel crunches beneath the wheels and my heels sink into the ground as I push her down the narrow pathway leading to my father's grave.

When we reach the family plot I unlatch the chain dangling across the entrance, then wheel my mother through. I wish I could spare her this. Surrounded by graves marked by weathered stones, my father's brand-new tombstone looks out of place, like a skyscraper erected in the midst of an old city.

My mother inhales sharply.

"Oh, Paul," she murmurs, reaching out a hand, letting it drop.

The rabbi begins singing and swaying over the grave as we gather in a semicircle. There are far fewer people here today than there were at my father's burial eight months ago. Lenny flies into my mind and buzzes around for a moment before flying back out. I try to conjure his face, but I can only see certain features: moist brown eyes, the side of his thick neck.

My mother is struggling to her feet.

"What are you doing?" I whisper, bending over her.

"I want my walker," she says.

"It's in the car—"

"Then help me stand," she insists.

I hold my mother's elbow, steadying her as she rises. She fishes for the pages in her jacket pocket, puts on her bifocals and begins to read. The words are a blur, vaporous as they float through the air. I watch her mouth move, I watch the pages flutter in the breeze.

*   .*   *

A few months from now, my father's brother will die suddenly of a heart attack. The smooth ground that now lies next to my father's grave will be dug up, and a plain pine box will be lowered into it. My mother will not be welcome at Harvey's funeral, but I will attend. I will stand off to the side, apart from Shirl, Moe, all the nieces and nephews. No one will speak to me. Even Susie will turn her back. As my mother's daughter, I have been tried and convicted for her sins. My chest will feel torn open and empty, and I will wonder whom I'm mourning. My uncle? My father? My mother? Myself?

But today—as I gather a few pebbles and rocks from the earth near my father's grave and place them on top of his tombstone—I tell myself that I'll try to hold on to them all. Surely Shirley and Harvey will understand that if it were up to me, it wouldn't have turned out this way. They'll realize that I have no choice but to stand behind my mother. After all, if I don't stand behind her, she'll fall.

"We're all going back to my house," my mother announces, folding the pages neatly into her pocket as I wheel her from the family plot.

"Of course we are," I say softly. "Of course."

I am walking up Madison Avenue on my way home from getting my hair cut when I see Jess moving toward me. I almost don't recognize her. She's dressed like a grown-up: suit, heels, her hair swinging just above her shoulders. She's carrying a briefcase. I, on the other hand, am a college student once again, wearing jeans and holding a dog-eared paperback of a nineteenth-century novel. By the time I realize it's Jess, it's too late to avoid her. I lower my eyes and hold my breath as we pass each other, and suddenly I feel a hand grabbing my arm, and the force of that hand shoving me hard into a mailbox on the corner. I whip my head around and stare at her, stunned. Passersby slow down and watch. Jess keeps walking, then turns around, still in motion.

"Whore," she tosses over her shoulder.

I hear people snickering.

"Jess!" I call after her. "Wait!" But she doesn't stop. I stand there, watching as her back recedes into the crowd.

I head home in a daze, the word *whore* ringing in my ears. I walk across the park, hating myself. She's right. I'll never live this down. No matter what I do, what I accomplish, for the rest of my life this will never go away. When I get to my apartment, I flop on the bed and stare at the ceiling. Jess looks as if she's gotten her life together—or rather, that she never let it fall apart the way I did. With her briefcase and sharp little suit, she might be a lawyer or an investment banker.

The phone rings, and I have an instinct that makes me answer it instead of letting the machine pick up. She doesn't even identify herself. It is the first time we've spoken since the night of her twenty-first birthday party, four years ago.

"You're through with him, aren't you?" she asks.

I sit up in bed.

"For a long time, now," I answer.

"I could tell when I saw you on the street. You looked healthy. Beautiful. You have color in your cheeks."

Her voice, lilting and musical, is as familiar as if we've suddenly slipped back to our Sarah Lawrence days.

"Why did you set me up like that at your party?" I ask. "I told you I didn't want to see Lenny again—why didn't you tell me he'd be there? What started that night came close to ruining my life."

I can hear her breathing. I don't want her to hang up.

"It's hard to explain," she answers slowly, "but I had my reasons."

"I never meant to hurt you," I say, as if my intentions could possibly matter. And for a moment I imagine that all things, even terrible, twisted things, can somehow be healed. "Listen, can I see you? Can we talk?"

"I don't think so," she says gently, almost as if she were trying to forgive me. "You've gotten rid of him. Be grateful for that. He's married to my mother, and he's the father of my little sisters. I'll never be rid of him for the rest of my life."

On the campus of Sarah Lawrence College, I keep my head down and carry my notebooks close to my chest. It's near the end of the second semester, and even though I'll be graduating in May, I still expect someone to tell me I have no right to be here. I spend most of my time alone in the library or in the cafeteria, drinking coffee and eating cookies. It has been eight months since my last drink, and one thing I've noticed is that I crave sugar all the time. I keep a big bag of plain M&Ms in my freezer and grab a handful every hour or so while I'm studying.

Ilja Wachs was true to his word: his nineteenth-century literature class has required about a thousand pages a week of reading. We've been doing the French recently, and I've been introduced to Balzac, Stendhal, Flaubert. But more important, I've discovered a way of reading and interpreting that is entirely new to me. I spend hours in conference with Ilja, hashing out ideas about how the metaphors in *Madame Bovary* might find their roots in the French culture of the time. Ilja has me reading a biography of Gustave Flaubert as well as some nineteenth-century French history.

In the meantime I'm writing. My creative writing class meets Wednesday afternoons, and today we're discussing the third short story I've written this semester, called "Sonata Quasi una Fantasia." I don't know enough to be embarrassed by the title. I have fallen in love with language, with the sounds of words, much in the same way, as a child, I heard the pattern of piano music in my head. I have been returning to the same character again and again: a young woman I've named Carolyn in my stories, who is based on Jess.

Jess has become a part of my internal landscape, an object of longing and obsession, and the only way I'm able to even begin to make sense of her is through the writing of fiction. I try to enter Jess's consciousness, to imagine what must have been happening inside her. I have so many questions; I examine each facet through my writing, as if I were holding a hard stone to the light.

In the workshop, I'm nervous as the students assemble around the large, round table, each holding a marked-up copy of my manuscript. My two earlier stories haven't gone over so well. They were too elliptical, just skimming the surface. So this time I've dug deeper, thought harder. I didn't want to turn this work in for the class to read, but the professor pushed me to do it. She's a published novelist—the first I've ever met—and I have an academic crush on her. She's a middle-aged woman with flowing black hair who wears kimonos and baggy pants. When she talks, her hands flutter by her sides.

"So what do we think?" she asks.

Silence.

I stare at my hands, which are folded in my lap so no one can see them shaking. I pick at a hangnail on my thumb until it bleeds.

"Anyone?"

Someone clears her throat, but no one says a word. It's all I can do to stay in my seat. I want to bolt out the door, run through the building and down the hill, all the way to the parking lot. I never should have turned in this story. It's too personal, too revealing.

I glance at the professor, and she gives me an encouraging smile, but she's not going to bail me out. It's her policy that the students speak first.

"Come on, people," she says, losing patience.

I'm about to cry, when finally Katie, the best writer in the class, starts to speak.

"It's beautiful," she says.

I think I've misheard her.

"That's why no one's saying anything. We don't know what to say."

The professor sits back as the students finally start speaking. Until now, I have been the quiet one in the class, too insecure and frightened to say a word. But today, something begins to shift. I see that there might be some way I can take the raw material of my life and transform it into something that transcends my own experience. I can organize the noise in my head into something that has order and structure. I can make sense of what, until now, has been senseless.

At the end of class, the professor pulls me aside. "There's a graduate writing program here," she says. "I've been meaning to talk to you about it."

I drive home from school in a state of dazed ecstasy. The moment the professor mentioned graduate school, I knew it was the right thing to do. One of the few gifts of spending so many years doing the wrong thing is the clarity with which I can see when something is right. Slowly, my life is gaining a sense of direction. By the time I walk into the lobby of my building, I'm breathless, euphoric. It's eight o'clock at night, and I have hours of work ahead of me: at least a hundred pages of *Anna Karenina* to read for tomorrow's seminar.

I say hi to Ernie, my doorman, as I head for the elevator.

"Hold on a minute, Dani, I've got something for you," Ernie says, rummaging in the drawer of his stand.

"Here it is." He hands me a small business card.

"This guy was here to see you. He asked you to call him as soon as you get in."

I look at the card: "Special Agent Richard Flagg, Federal Bureau of Investigation."

My heart starts to pound. There must be some mistake.

"Did he say what he wanted, Ernie?" I ask.

"No, but he did start asking some questions."

"Like what?"

I feel as if I'm hallucinating.

"Oh, how long you've lived here, and whether you live alone—stuff like that."

Ernie is looking at me with a curiosity indigenous to New York City doormen. He knows there's a juicy story here.

I walk to the elevator. I feel Ernie's eyes on my back. It must be a mistake—it must! What could the FBI possibly want with me? I've never even gotten a speeding ticket.

As soon as I unlock the door to my apartment, I flick on the lights and go straight to the phone. I don't even take off my jacket. I sit on the edge of my rocking chair and dial the number on the card.

"Flagg," a voice barks.

"This is Dani Shapiro," I say as evenly as I can. Why am I so nervous? I haven't done anything wrong.

"Hi, Dani," he responds, as if he knows me.

I don't say anything.

"Do you know what this is about?" he asks.

"I have no idea," I answer.

"None at all?"

"No," I say. I'm beginning to get pissed. Maybe this is a crank. Maybe I should call a lawyer.

"Lenny," he says.

A single name, and the world disappears.

CHAPTER
NINE

"Don't talk to them," my lawyer snaps when I call her the next morning. "Let them subpoena you."

"Subpoena?" I repeat. What the hell has Lenny done? And what does the FBI want with me?

"How did they find me?" I ask.

My lawyer laughs.

"Give me a break," she says.

My lawyer, like every other lawyer in America, is aware of Lenny Klein. In the past couple of years he has become notorious for being responsible for the demise of the largest firm in New York. His raccoon coat and white Rolls-Royce have been written about in the press, emblematic of the high-flying eighties, which are about to be over. There is nothing Lenny feels he can't get away with. This is the kind of news that, in the past, would have made me run for the nearest drink.

"This really scares me," I tell my lawyer.

"I'll call them and try to get the lowdown," she says.

*   *   *

When I hang up the phone, I try to get some work done. I've begun to turn those short stories into a coming-of-age novel about a young Orthodox Jewish girl—but I can't concentrate. Ever since my father's death, I've felt like a convalescent: I go to AA meetings almost every day, therapy twice a week, sleep nine hours a night—a hard, dreamless sleep from which I am jolted by the alarm clock's early-morning ring. I feel constantly raw, as if the years of drinking coated my nerve endings, and now that the coating is gone I feel everything: fear, panic, rage, anxiety, all in one big stew. I find it unbearable. It seems to me that my emotions have the power to kill me, that they will seize me by the throat and never let me go.

I'm in the second year of the master's program at Sarah Lawrence, and I've just turned in a long section of my novel. It seems the class is having a problem finding the Lenny character believable. I've written him as he is (or at least how I think he is) and the other students keep telling me that he just doesn't ring true: no young girl with as much going for her as my narrator does would ever find herself in the thrall of such an unappealing man. *At least make him more physically attractive,* someone said. *Yeah, or make the narrator more of a loser so we understand why she's with him.*

I stare at my computer screen.

The phone rings. Usually I turn off the ringer when I'm working, but today real life, or at least real history, is intruding.

"Okay," my lawyer says. "You're not going to believe this."

She tells me that the U.S. attorney's office is seeking grand jury indictments against Lenny for various counts of racketeering, conspiracy, mail fraud, and obstruction of justice. Apparently Lenny has lied not only to me but also to his clients, such as Shearson Lehman and Home Insurance. Even Donald

Trump is angry at Lenny. I feel a weird surge of joy at this news, because for an instant it makes me feel less like an idiot for believing Lenny's lies for as long as I did.

"So I think they're going to subpoena you," my lawyer is saying.

The room swims. All my life I've felt as if I've done something wrong. When I see a police car on the highway, even if I'm not speeding I think I'll be pulled over. And when walking through metal detectors in airports, I imagine that the alarm will sound and a gun or a bomb will be pulled out of my luggage.

"Am I in trouble?" I ask.

"I don't think so," she says. "They just want to talk to you—find out if you know anything."

"Why?"

"They won't say, exactly. But let me ask you this, Dani. Do you know anything you want to tell me right now?"

"What do you mean?"

"Did you have any idea that this was going on?"

"No!" I blurt out. "Of course not!"

"Don't get upset," my lawyer soothes. "I just had to ask."

In the two years since my father's death, the following people have died: Harvey dropped dead of a heart attack; Hy died of pancreatic cancer; my paternal grandmother finally expired after twenty bedridden years; Susie's mother died of breast cancer. I have grown accustomed to funerals, burials, weeks of sitting shiva. I know the way to the Brooklyn cemetery, and no longer get lost when walking through the maze of tombstones, looking for the family plot. Sometimes I drive there and just sit on the low stone bench for a while, surrounded by new graves. I keep telling myself this will end, that the odds have it that the people I love will stop dying so rapidly—that a phone call in the middle of the night will not always bring bad news.

In the midst of it all, I am trying to build a life for myself, but my mother isn't making it easy. She graduated from wheelchair to walker to cane, and now strides down Broadway, arms swinging, a miracle of medical science. The only clue that her legs were shattered is the fact that she has to take one step at a time when climbing stairs. She looks a decade younger—she is sixty-five—and she is incandescent, lit from within by a rage she has carried all her life, and which, at the moment of the crash, became her life source.

My mother's rage has been her ticket to survival. She has used it to lift her arm weights, do her leg exercises, fight legal battles with Susie, Shirley, Harvey, and the insurance company. She has even contemplated a medical malpractice suit against my father's internist for not foreseeing that he would pass out behind the wheel. The cause of the accident, no matter how it's picked over, will never become clear. Did my father have a stroke that caused him to pass out at the wheel? An arrhythmia that could have been prevented? Had he taken one too many tranquilizers that day? Now that her legs and arms are strengthening and her lawsuits are being settled one by one, my mother is letting go of some of these questions, and has turned her attention to me.

"Why haven't you called?" she asks, two days after we had lunch. "My friends can't believe it. A daughter who doesn't call."

"Mom—"

"Am I being paranoid? Do you not want to spend time with me?"

"I'm busy with schoolwork," I tell her. "I'm trying to write this novel."

"And that means you can't find five minutes to talk to your mother?"

Her voice, over the phone line, is stretched tight. She has accepted this as her due. Children take care of parents. Sacrifice is the name of the game. She was middle-aged before she had to take care of her own ailing mother, but still she compares her situation to mine. *I went to the old-age home every week*, she says to me. *And I did it cheerfully.*

A little knot inside my head unravels.

"You know, we saw each other Tuesday," I say. "We had lunch."

"Tuesday! Today is Thursday and I haven't heard a word from you," she responds triumphantly, as if I've just made her point for her.

"Mom, you may find this hard to believe, but a lot of my friends only talk to their parents once a week, if that," I say.

Don't try to explain, I tell myself. It's not worth it.

"And *my* friends tell me their grown children invite them to spend whole weekends at their houses," she says, referring, I can only assume, to people a decade older than myself, who live in Scarsdale and have already produced grandchildren.

"I'm sorry I'm not the daughter you wish you had," I snap.

There is silence on the other end of the phone. I twist the wire hard around one finger like a tourniquet.

"How dare you!" my mother shrieks. "I almost died!"

Now rage is bubbling through my own veins.

"Give me some room!" I shout at my mother. "I'm just trying to put my life together!"

"At least you *have* a life," she cries. "I have nothing."

I close my eyes and my hand goes slack around the receiver. She's done it again—pulled out the big one, the nuclear bomb in her arsenal. The words echo in my ears—*I have nothing*—and I feel them enter me, twisting themselves around my ribs until they are a part of my very bones.

"I'm sorry," I whisper.

I am a terrible daughter and deserve to be punished, I think. My mother has nothing, and I have so much.

My mother doesn't say a word.

"I'll do better," I say.

Still nothing.

"Do you want to have coffee this afternoon?" I ask. "I'll come by around three."

"Sure," she says, and I can almost see her shrug. "If you'd like."

My lawyer's office on lower Fifth Avenue is located in the same building as the offices of a casting director I used to regularly audition for, and when the elevator door opens to reveal a bench of cross-legged blondes wearing miniskirts and reading television commercial scripts, for a split second I think about getting off the elevator and blending in. But then I remember: I have gained ten pounds since starting graduate school, and just last week I cut all my hair off.

The elevator door slides shut and my heart skips a beat, reminding me of the exact nature of the trouble I'm in. Upstairs, in my lawyer's office, two FBI agents are waiting for me, and they want me to tell them everything I know about Lenny Klein.

My lawyer is waiting for me in the receptionist's area. The fact that I have a lawyer at the age of twenty-five is a direct result of what has happened in my family: Susie suing my mother, my mother countersuing Susie, my mother trying to remove Harv and Shirl from their position overseeing a family trust, my mother trying to have my grandmother declared incompetent.

And now this.

"Hi, kiddo," she says, kissing me on the cheek.

She smells like rose water and oatmeal. She's a wife and a mom and a lawyer—someone who would never be called into question by the FBI. She's wearing a knee-length skirt and shoes with stacked, sensible heels, a gold wedding ring. Her black hair, flecked with gray, is brushed back from her face in a simple, short style. I would do anything right now to switch lives with her.

"They're waiting in the conference room," she says.

"How do they seem?"

"Like feds," she says, as if that might provide me with a mental picture.

She ushers me into the conference room. I see myself reflected in the glass door. I look like the graduate student I am: jeans, boots, sweatshirt, newly cropped hair. Whatever a rich guy's mistress looks like, I am no longer it.

"This is Dani Shapiro," she announces, and two men in dark suits swivel around in their seats to look at me.

"Agent Flagg"—one of them shakes my hand limply—"we spoke on the phone."

I nod once—what I hope is a don't-fuck-with-me nod—and sit.

"And this here's Agent Anderson." He gestures to the guy next to him.

I nod again. I am determined to speak only if spoken to.

"Coffee?" my lawyer asks.

"Thanks," replies Anderson, even though she wasn't addressing him.

Flagg folds his fingers under his chin, elbows on the conference table.

"So."

That's all he says. So. Just like that. As if I'm going to sit here and start talking. He has a yellow legal pad in front of him, and a Bic pen poised in his fingers. A microcassette recorder is next to the pad, spools turning.

I raise my eyebrows and meet his gaze. I probably look cool and calm to him, but if he could see what's going on inside me, he'd see another story. I am buffeted by anger and shame; anger that I have to be here, that Lenny Klein has infiltrated my new life—and horrible shame that I ever was with him.

"Let's get on with it, gentlemen," my lawyer says crisply.

Flagg glances down at the blank page on his legal pad, then back up to me.

"Lenny," he says, "is in big trouble."

The way he says it, *Lenny*, the way his mouth forms the syllables, is almost lewd. It connotes an intimacy of sorts—or at least a knowledge that he's dealing with someone who has been on intimate terms with Lenny Klein.

"What does that have to do with my client?" asks my lawyer.

"As I explained to you over the phone, your client hasn't done anything wrong," Flagg says. "We just want to see if maybe she knows something she doesn't even know she knows."

"Like what?" I blurt out. This is getting to be too much for me. I feel as if I've stumbled onto the set of the wrong movie.

"For instance, you and Lenny traveled," Anderson says in a statement, not a question.

"Yes," I reply faintly.

"Did you travel within the continental United States?"

"Yes."

"Did he purchase your tickets for you?"

"Sometimes."

"Were they always under your own name?"

The questions are coming faster and faster. I feel dizzy. I don't understand the significance of this line of questioning. But

I promised myself before I got here that I would answer any questions honestly.

I think about whether Lenny always bought me tickets in my own name. When we flew to Europe I used my passport. But what about other trips? Suddenly, I flash on trips to LA, and the names of various associates in his law firm on tickets I used.

"No," I answer slowly.

A shadow of a smile crosses Flagg's face.

"And whose names were they in?"

"Associates of his," I answer.

"Bingo," murmurs Anderson.

"I don't get it." I turn to my lawyer.

"He was charging clients for your airfare," she responds.

I take a deep breath. I remember the way Lenny used to hold our tickets as we were boarding—these were the days before airlines asked to see identification for domestic travel—and how we would often check into hotel rooms under the names of his associates or even sometimes clients. To the extent that I was aware of this at all, I imagined it was so that his wife wouldn't be able to find us. It never occurred to me that he might be doing something illegal.

"I didn't know," I mumble.

"Okay." Flagg squints out the window at the building across the street. "Let's move on."

"What about gifts?" Anderson asks. "Lenny buy you any big items?"

"Yes," I say softly.

"Such as?"

"Jewelry, furs, cars," I answer. My resolve to look him in the eye is fading fast.

"Jewelry, furs, cars," Anderson repeats like a mantra. He seems to be stifling a laugh. "Would you care to elaborate?"

"A mink coat," I say, "a watch, a diamond ring—"

"A diamond ring? Were you engaged?"

"Hardly," I say, and I hear an edge creeping into my voice.

"Keep going," says Flagg. "Any other jewelry?"

"Pearls," I say. "He bought me a strand of pearls in San Francisco."

"And what would you say is the approximate value of all the jewelry?" asks Anderson.

I make a mental tally. Lenny always let me know the prices of the things he bought me. That was part of the game. He enjoyed my knowing how much money he was spending on me. I used to tell him to stop buying me things, that I didn't want or need more jewelry and that I felt silly wearing fur coats, but he just kept going. After all, what was a mistress for, if not to drape in luxury like a mannequin?

"I guess about fifty thousand dollars," I say.

Anderson lets out a low whistle.

"How did he pay for these items?" he asks. "Check? Plastic?"

"Plastic, usually."

"Do you remember if he used his own personal card or an expense account card?" he asks.

"I don't know. Probably both," I say.

"And what about the vehicle he purchased for you?" he asks.

"First he leased me a Ferrari," I say. "Then he bought me an Alfa Romeo"—I pause for a second, flashing on how on earth I could have ignored what was so obvious—"then a Mercedes."

"Sedan or convertible?" Flagg asks.

"What difference could that possibly make?" my lawyer snaps.

"Just asking," he shrugs.

I stare out the glass door of the conference room, flooded with self-hatred. These men think I was with Lenny for his money—of course they do. Their follow-the-money detective-school training allows for no other possibility. I feel like telling them that if I had been with Lenny for his money, I would have let him buy me the million-dollar apartment he begged me to take on Central Park. *I thought I was in love with him,* I want to tell them, but why would they believe me?

"Did you keep these items when you ended your relationship?" Anderson asks.

"Some of them," I say, omitting the fact that I sold the Mercedes to pay for graduate school.

"Has Lenny been in touch with you?" Flagg asks.

"He's called, but I haven't spoken with him," I say.

"When was the last time?"

"About three months ago," I say.

Anderson turns to Flagg.

"It's interesting," he murmurs. "The pattern is quite different than with the other girls."

"Other girls?" I repeat.

"Oh sure," he says.

"So are we about wrapped up here?" my lawyer interjects.

"What other girls?" I ask.

For the rest of my life, Lenny Klein will be an eternal question I will try to answer. I will struggle to accept the fact that I spent nearly four years—four *formative* years—with a man I never knew.

"Six," Flagg says without a blink. "There were six, including you."

"Have you talked to all of them, or are you planning to torture only my client?" my lawyer asks.

"Oh, we're talking to everybody," answers Anderson. "But with Dani, here, it seems Lenny was more . . . *open* . . . than usual. He seemed to almost be trying to set up a life with her. The others were more like . . . the usual fare."

"You know," Flagg says, "you should be prepared for some phone calls from the press when this breaks."

"What the hell do you mean?" my lawyer snaps.

"There have been some leaks," Anderson says. "Some of the other girls have already been contacted. A few of them have agreed to appear on news shows."

"And where did those leaks come from?" my lawyer asks. "You guys are disgusting."

With this, Anderson and Flagg begin to pack away their apparatus.

"We may need to speak with you again," Flagg says.

I glance pleadingly at my lawyer. Will this never be over? I feel as if I've clawed my way out of a pitch-dark pit only to reach the top and have my fingers pried off, one by one. I am flailing, falling. Jess was wrong—I will never be able to escape. There's no way out.

"Give her a break, guys," she tells them.

"We'll try not to call her to testify," Flagg responds. "But I can't make any promises."

Years later, I am at a party filled with famous writers: a beautiful Irish woman, a burly, hard-hitting reporter, a young guy who just got a big advance for his new novel. I am sure no one knows who I am, and I am sipping seltzer near the bar, making conversation with the bartender, whom I know from AA, when a man I recognize approaches me.

"Are you Dani Shapiro?" he asks.

I can't believe he knows who I am. He is a well-known journalist who regularly pops up in the society pages, and I'm a young writer with one book published.

I nod and shake his hand.

"We know someone in common," he says.

"Who's that?" I ask with a smile. I am flattered, delighted.

"Lenny Klein," he says.

His inquisitive gaze focuses itself, laserlike, on my expression as he watches me blanch.

"He's no friend of mine," I manage to stammer.

"Oh, really? I'm writing a piece on the trial for *Esquire*," says the journalist. "I thought you knew him pretty well."

It has been four years since I've seen Lenny Klein, almost two years since the FBI subpoenaed me. My lawyer has warned me of the possibility of paparazzi outside my building, and phone calls from reporters have been left unreturned. I have narrowly averted becoming a sideshow in Lenny's very public trial. And now, sipping seltzer in a book-lined parlor, here I am again.

"I don't want to talk about this," I say. Beneath my sundress I am turning red and blotchy, heat spreading up my neck and across my face.

"He's going to jail, you know," says the journalist. His breath smells like white wine gone sour. "He was sentenced this morning."

I walk away and feel his eyes on my back.

"Have a nice day!" he calls after me.

I stride through the parlor, out the front door of the brownstone and down the steps. I try to imagine Lenny in jail. Will they let him keep his hairweave? Will his wife make conjugal visits? Will his kids still be able to go to prep school?

I occupy myself with questions of this nature; they flip quickly through my mind like flash cards. I walk faster, until I practically break into a run. A man I spent four years of my life with has been convicted and sent to jail, and I feel empty inside. I might feel a surge of sick joy or I might pity him. Instead, I feel nothing but relief that it's over.

The sun is unseasonably strong on graduation day. It beats against the yellow-and-green-striped tent pitched on the Westlands Lawn of Sarah Lawrence. The air is humid and still, and bees hover low to the ground. My hair is sticking to the back of my neck, and I feel sweat gathering in the creases behind my knees as I sit with the dozen other graduate students waiting to receive our Master of Fine Arts degrees.

As I listen to the keynote address my mind drifts to my mother, who is in the audience. For her—for both of us, really—this day is about my father's absence. Could he have imagined this for me? Would he have chosen it for me, if he had a choice in the matter? Six weeks ago, the novel I began while still in college was sold to a major publisher. I just gave my first interview, in which I was called precocious at the age of twenty-seven. When I read the piece, I barely recognized myself. It's been a long time since I've felt young or precocious.

The president of Sarah Lawrence is calling out my name and I rise, my heels sinking into the soft, damp earth as I walk to the podium and shake her hand. There is a smattering of applause from the few friends I have made here. I'm sure some of my classmates aren't clapping at all. I have done the unforgivable: secured a publishing contract before graduation. *If you're going to envy me, envy all of me,* I want to say to them. *Envy losing half your family in a year.* Clutching my diploma, I scan the crowd for my mother. She is sitting on the aisle about halfway back, wearing a white blazer and sunglasses. She is applauding wildly—I could swear she shouts *Bravo!*—and her brother, Morton, is sitting next to her, holding a video camera. She leans over and whispers something in his ear.

Eza bat yesh lee, I imagine she's saying.

* * *

After the ceremony there's a reception on the great lawn. I see Professor Ilja Wachs across the way, chatting with some parents, and wonder if he has any idea of the hand he had in changing my life for the better. My first writing teacher ambles over and gives me a kiss on the cheek.

"You did it," she whispers. "Now you just have to keep doing it." She tells it to me as if it were an anagram, a message it will take me a lifetime to decipher: doing it and keeping on doing it are two different things. I will never be out of the woods.

"Dan-o," Morton finds me and crushes me into a hug, grinning the megawatt grin indigenous to my mother and her siblings. "Whatta girl."

My mother holds me at arm's length, staring intensely into my eyes. I don't know what she thinks she sees there. Now that there is no question about my mother's recovery, now that she is solidly on her own two feet, I find myself mired in our shared history. Why didn't she try to stop me from being with Lenny? Why didn't she intervene somehow, when I weighed one hundred pounds and lived on white wine and vodka? Sometimes I want to hurl at her the very same accusation she hurls at me: *I almost died*, I want to cry accusingly. *Where were you?*

Now, she kisses the tip of my nose.

"My beautiful daughter," she announces.

I try to smile, and we hug stiffly. I feel the bones of her back, the new strength in her arms. I am her only child, and she has survived for me. The least I can do is spend the rest of my life trying to make her happy. What she doesn't realize is that I have survived for her as well—and only now am I beginning to survive for myself.

"Gotta go," I mumble. I know I should stick around and let

my mother *kvell* with motherly pride on my graduation day, but I just can't take another minute.

"Have lunch with us, darling," she says.

"Let's make it dinner," I say, unbuttoning my graduation gown.

I throw my diploma into the trunk of my car and head back to the city. I blare Bonnie Raitt on the tape deck and keep the windows rolled down as I pass through two tolls and onto the West Side Highway. The George Washington Bridge is suspended across the Hudson River, and whitecaps sparkle in the midday sun. The curves of the Upper West Side unfurl like a banner.

Instead of getting off the highway at the Seventy-ninth Street boat basin, I keep going. I turn off Bonnie Raitt and listen, instead, to the strains of a Hebrew melody so familiar I might have heard it in the womb. My father's voice drifts through me—loud and off-key—singing "Havdalah," the song that marks the end of Shabbos. I see his open mouth, smell the spices as he shakes the ceremonial spice box, hot wax dripping from the braided candle onto the kitchen counter. I feel his hand resting on my shoulder. I should pull off the road—tears are blurring my eyes—but I keep driving downtown. I don't know where I'm going until I'm halfway there.

There are rumors of wild dogs who roam the cemetery. I usually carry a baseball bat when I visit, but today I am empty-handed. I park my car as close to the family plot as I can, unhook the chain, and sink to my knees at the foot of my father's grave. I know he's not here. Here, there is only a tangle of bones. But still, I can talk to him in this place without second-guessing myself, without wondering if he really hears me.

"I wrote a book, Dad," I whisper at the ground.

No kidding, I can almost hear him say.

"I just got a graduate degree."

How about that.

"I'll be good to Mom."

I know you will.

The sun is hot against the back of my bowed head, and my bare knees are damp and grass-stained. I am surrounded by the remains of my father's family, here in the bowels of Brooklyn with the el train rumbling overhead.

"I'm sorry," I whisper. "You never got to be proud of me."

Silence. I have lost his voice, and his face is fading from my memory, slowly, like an old photograph. A bird pecks at the earth near his grave. I stand up, brush myself off, and look around the family plot for loose rocks to place on top of each headstone. I place pebbles on my grandfather's grave, then my grandmother's, then Harvey's. Finally, I take the biggest one I can find and rest it above my father's name.

ACKNOWLEDGMENTS

This book would never have come into being if not for the goodness of the people who were there for me when I was not able to be there for myself: Anstiss Agnew, the late Jerome Badanes, Loretta Barrett, HGWB, Diane Brandis, E. M. Broner, Rob Brownstein, Annette Gartrell, Roberta Golubock, Elizabeth Morris Krinick, the late Alexander Lipsky, Emily Klein McGrath, Rollene Saal, Laura Van Wormer, Ilja Wachs, and friends of Bill W. everywhere.

Jill Bialosky, Betsy Carter, Jill Ciment, Elizabeth Fagan, Donna Masini, Jan Meissner, Mary Morris, Helen Schulman, and Melanie Thernstrom read early drafts and offered invaluable insights.

I thank Dr. Richard Zimmer for his talent and wisdom.

My family deserves a special debt of gratitude: it can't be easy to have a writer in their midst—particularly one who has written about them.

Thanks to the Corporation of Yaddo, where a portion of this book was written.

Thanks also to Deb Futter, André Bernard, and my wonderful agent, Joy Harris.

Most of all, my husband, Michael Maren, helped me feel safe enough, read every word, and made me believe in happily ever after.

About the author

About the book

Read on

Insights,
Interviews
& More...

A Conversation with Dani Shapiro

The following conversation appears in "Meet the Writer," an exclusive interview with Dani Shapiro conducted by BN.com. Reprinted by permission of Barnes & Noble.

What was the book that most influenced your life or your career as a writer?

When I was in graduate school, I took a course in nineteenth-century literature that changed my life as a writer, in the sense that it changed my life as a reader. I had a gifted professor who taught me to read very differently than I had before. Until then, I had read novels through a somewhat academic lens. But in this course, the professor (whose name was Ilja Wachs, and to whom I eventually dedicated my second novel) taught us to read as *writers*. It was as if a lightbulb had gone off in my head. For the first time, I began to understand that metaphor, simile, foreshadowing, and such were part of the creative process— that the writer wasn't necessarily maneuvering and making decisions as much as following unconscious motivations. The book I was reading at the time the lightbulb went off was *Madame Bovary* by Gustave Flaubert. I read it three or four times, and it also led me to read about nineteenth-century French history in order to understand the political and social factors that might have been influencing Flaubert—since writers cannot help but be affected by the times in which they live.

What are your ten favorite books, and what makes them special to you?

What a hard question. This list would probably be somewhat different on any given day. I'll begin with older, classic books and make my way toward the contemporary ones:

Daniel Deronda by George Eliot— I read this in the same nineteenth-century literature class. It deals with ethnicity, with Jewishness, in a way that was highly unusual and risky for a writer of that time and place—particularly a female writer.

Mrs. Dalloway by Virginia Woolf— I could really list any novel by Woolf. Reading her prose is like looking through calm, clear water. The language is so perfect. It does what great writing should: it provides illumination without ever getting in the way.

Lolita by Vladimir Nabokov— Gorgeous, poetic, verging-on-mad prose. An orgy of prose, really. I read it for the first time at an artist's colony where I was visiting while writing my first novel, and I remember feeling sad, almost nostalgic while reading it because I'd never be able to read it for the first time ever again.

A Fan's Notes by Frederick Exley— When it was published, it was billed as a "fictional memoir." I've taught it for years in a course called "The Literature of Autobiography." It's a fascinating hybrid which raises interesting questions about truth and fact and fiction. What is the ▶

66 Reading [Woolf's] prose is like looking through calm, clear water. 99

A Conversation with Dani Shapiro
(continued)

difference between autobiographical fiction and memoir? *A Fan's Notes* is provocative, moving, painful, and always beautifully written. It was Exley's great triumph, and that triumph is the redemption within the book itself.

Sleepless Nights **by Elizabeth Hardwick**—This long-out-of-print book is one of the great coming-of-age stories set in New York, by one of our finest writers. It still feels very modern to me, and is the most classic and powerful of the whole single-girl-in-New-York genre.

Slouching Towards Bethlehem **by Joan Didion**—I love everything Didion has ever written. I'm amazed by the way she has this razor-sharp intelligence which doesn't interfere with—in fact enhances—the beauty of her prose. I could read a sentence from her work, taken completely out of context, and know it was hers.

Revolutionary Road **by Richard Yates**—The great suburban novel. Yates is uncompromising, unblinking about his characters, and yet he maintains at all times complete sympathy for them. A heartbreaking book.

Cat's Eye **by Margaret Atwood**—A beautiful novel about girlhood friendship—a subject which fascinates me. My favorite of all of Atwood's novels.

The Furies **by Janet Hobhouse**—I buy copies of this book whenever I come across them. Hobhouse was stricken with ovarian cancer at forty-two as she was writing *The Furies*. Her awareness of time running out

66 I could read a sentence from [Didion's] work, taken completely out of context, and know it was hers. 99

gives the book a powerful sense of momentum. It's an autobiographical novel, gorgeously written.

Patrimony by Philip Roth—A sad, touching, funny, honest memoir about Roth's father. I teach this book as an example of how a memoir can be about one specific relationship, rather than a kitchen sink of life. I love Roth—all of Roth—but somehow this is the book of his that resonates with me most deeply.

Do you have any special writing rituals? For example, what do you have on your desk when you're writing?

I need a lot of quiet, and very much the sense that I'm in a room of my own. And coffee. Good coffee is very important. I've begun, in the last several years, to write longhand, and then eventually transpose into my computer. I carry a notebook around with me. I have these special notebooks that they only sell in the bookstore of my husband's hometown. So whenever my in-laws come to visit, I beg them to bring me more notebooks!

Many writers are hardly "overnight success" stories. How long did it take for you to get where you are today? Any rejection-slip horror stories or inspirational anecdotes?

I sold my first novel while still in graduate school, so in that sense my ▶

“ I teach [*Patrimony* by Philip Roth] as an example of how a memoir can be about one specific relationship, rather than a kitchen sink of life. ”

A Conversation with Dani Shapiro
(continued)

career as a writer began without a whole lot of rejection. But aside from that, my writing life has been all about slow, steady progress. All I ask of myself is that I get better with each book. And I think I have. Hopefully I always will.

What tips or advice do you have for writers still looking to be discovered?

Be true to yourself. Do the work that matters to you, and don't think for a single second about the marketplace and what publishers might or might not be looking for. I once had a student who studied the bestseller list and tried to write a bestseller. The work was awful. Once she turned to what truly mattered to her, she wrote a lovely first novel that was published and published well. Self-consciousness is the enemy of the writer.

> 66 Do the work that matters to you, and don't think for a single second about the marketplace. 99

The Story Behind
Slow Motion

The following essay first appeared in the Los Angeles Times

IN THE EARLY STAGES of writing *Slow Motion*, I packed my bags and prepared to spend a month at Yaddo, an artists' community in Saratoga Springs, New York. I had never been to Yaddo before, and was feeling intimidated: James Baldwin, Truman Capote, Sylvia Plath had been guests there. A composer I knew had told me a story about his first visit: the man helping him with his luggage brought him to a music studio which overlooked a crystalline lake. "Aaron Copland composed *Appalachian Spring* here," the man called over his shoulder as he left. "Best of luck!"

As I packed, I wondered what I might bring—what talismans or photographs—to help me remember the story I wanted to tell. It had been tough going. As a novelist, I felt at sea when it came to writing memoir. I was accustomed to my imagination leading the way. But now *I* had to lead the way. A journalist friend had suggested to me that I outline the story. "After all," she said, "you know what happened." But did I? Did I really know what happened? What did it even mean, to know what happened? I had been at it for a year, and I barely had a sentence down on paper.

It was then that I remembered my journals. Volumes of them, stuffed into the bottom drawer of an old file ▶

cabinet. The journals! I had kept them in various forms: pretty, fabric covered notebooks; simple lined ledgers; reams of typing paper from the years I started writing them on my computer. I ran over to the file cabinet and pulled at the bottom drawer.

It was locked. I yanked and pried. I stuck a paper clip into the lock and tried to jiggle it open. I kicked it, succeeding only in stubbing my toe. The harder I struggled to open the thing, the more determined I became. The locked file cabinet became a metaphor for the memoir itself. It seemed the success or failure of it—the very creation of it—hung in the balance. *What happened* was in there. Times, dates, even the weather existed on the pages of those journals. What did I have for breakfast the morning that I received the phone call that my parents had been in a terrible accident? What shoes had I worn to my father's funeral? What details of my life had I forgotten? Finally—this took hours— I broke open the drawer with a metal ruler, leaving a dented, useless file cabinet behind.

The next morning, I packed my journals in the trunk of my car and drove to Saratoga Springs. I was shown to my room—a grand, octagonal affair with more than a dozen windows and a fainting couch ("Carson McCullers wrote *The Member of the Wedding* here!") and then there I was. My baggage of every sort surrounded me: duffel bags, book bags, my journals, my computer. I had a whole month ahead of me. I unpacked in high spirits.

The next morning, I climbed up to my room after breakfast and settled onto the fainting couch. I had organized the journals the night before. At the top of the pile, the first red-and-white flowered one, which I had begun at sixteen. At the bottom of the pile, in boxes, the last typed pages from my late twenties. I had stopped keeping regular journals a few years earlier; it wasn't so much a decision as a sign that I had become . . . happier. More at peace. I found it less necessary to document the ins and outs, the ups and downs of my tumultuous personal life, because that personal life had become less tumultuous.

I opened the first one. The tiny, neat, girlish handwriting! Those hopeful loops! I had stopped just short of dotting my *i*'s with hearts, it seemed. As I began to read, a strange sense of disconnectedness settled over me. I wondered if it was just the

Yaddo effect. I had heard there were ghosts in the old mansion. Maybe Carson McCullers didn't approve of memoir as an art form. I kept reading, but with a growing unease. I put down the red-and-white flowered journal and picked up the next one. A thick, heavy journal with illustrations on each page, it was the record I kept during the year of my parents' accident. Maybe *this* was where I needed to begin.

I balanced the journal on my stomach. The girlish, loopy handwriting had been replaced by an undisciplined scrawl. I often had written late at night, and often, I was tipsy. It showed—both in the sloppiness of my handwriting and the unexamined nature of the content. It was shallow. It was uninteresting. I had written copiously about my feelings with no insight, no perceptiveness. I pressed on in horror. *Who was this girl? What had she been thinking?*

At some point during the reading, I must have fallen asleep. I'm not—nor have I ever been—a napper. Granted, I was lying on a fainting couch, but still I had no intention of drifting off into the thick, foggy, dreamless sleep that overtook me that summer morning. I slept for hours. When I awoke, I felt drugged. Poisoned, almost. I stared at the ceiling fan, its blades circling. It was then that it came to me: *my younger self wasn't worth writing about.* That girl with the loopy handwriting, that young woman whose banal, unexamined misery filled the pages of the journal still resting on my stomach—she was nobody's heroine. She couldn't possibly be the center of the story I was trying to tell.

I had thought I needed to get closer to the girl I had been. Now, I realized, if I spent one more minute in her company, I might abandon the whole idea of writing a memoir. I might flee the Yaddo mansion in shame. I packed up the journals into a big cardboard box and pushed the box into the deepest corner of a closet. I needed to forget them, I realized—to forget that they even existed.

Slowly, over that next month, I began to write my book in earnest, which is to say that I began to sift through my memory to find the shape of the story. This remembering was a delicate alchemy, part archaeology, part forensics, and—perhaps the most important part—a powerful creative urge to take that time in ▶

my life, those ashes, that sadness and self-destruction, and turn it into something larger and universal. To find the narrative in the tragedy. To make art out of loss.

But what was this art? I discovered that memoir is not a document of fact. It isn't a linear narrative of what happened so much as a document of the moment in which it is written. The present moment acts almost as a transparency, an overlay resting on top of the writer's history. The interplay of these two planes—the present and the past, the *me now* and the *me then*—creates the narrative and the voice. One can't exist without the other.

I was in my early thirties when I set out to write *Slow Motion*. My reasons for turning to memoir were, at least in part, writerly ones. I felt stuck in my fiction—haunted by the story of my parents' accident, which defined me at that time. My three novels all revolved around a central calamity. I felt like my own autobiographical material was controlling me. It was clear that I needed to wrestle my past to the ground. I needed to pin it in time, to capture it as if it were a wild animal that I could domesticate—or at least put behind bars.

But in the two years that it took me to finish a first draft, other things had begun to define me as well. I met the man I would marry. By the time the book was published, I was pregnant with my son. My life began to change. The tragedy of my parents' accident and the calamity of my own adolescent rebellion receded a bit, and began to feel like pieces of my past, rather than of my present.

If I hadn't written *Slow Motion* at the precise moment I did, it would have been a very different book. How would motherhood have affected the story? Or marital happiness? And if I were to have written the book today—at fortysomething? How would my mother's death have changed the writing of it? After all, it had changed *me*. A disturbing thought: Would I have felt the need to write it at all? Of course these are unanswerable questions. Ultimately, *Slow Motion* stands as a document of a particular moment in this writer's life—the only moment in which it could have been written.

It's a beautiful and strange thing that our lives shape our

memories, and our memories shape our lives. All we have—our entire consciousness—is memory. Events turn into memories even as they are happening—and memories become the stories we tell, both to ourselves and to others. What we are left with is the shifting nature of these memories, these stories. We stand from wherever it is we stand, and from our perch—at thirty, fifty, seventy, ninety—we see what we see. As painters use color and shape to imply closeness or distance, our memories, too, supply ever-changing color and shape. That white smudge in the distance might once have loomed large as a mountain.

As it happened, I did remember what I had been eating for breakfast the morning I received the phone call about my parents' accident: a half-grapefruit sweetened with honey. And I did remember the shoes I wore to my father's funeral: black heels I never wore again. But I didn't need those details to write a memoir. Nor did I need the words of that sad, lost girl I had been. The story, I discovered, wasn't about what happened. It was about layers of time, one on top of the next: a dialogue, a dance between two selves. *Me now, me then.* ∾

Have You Read?
More by Dani Shapiro

DEVOTION
A new memoir by Dani Shapiro

"I was immensely moved by this elegant
book, which reminded me all over
again that all of us—at some point or
another—must buck up our courage
and face down the big spiritual
questions of life, death, love, loss, and
surrender. Dani Shapiro probes all
those questions gracefully and honestly,
avoiding overly simple conclusions,
while steadfastly exploring her own
complicated relationship to faith and
doubt."
>   —Elizabeth Gilbert,
>   author of *Eat, Pray, Love*

"Dani Shapiro's novels and nonfiction
are always rich in honesty and
intelligence, about the psyche and lost
hearts and families, about messes and
shame and what calls us to transcend;
and how painfully we find out who we
are, and how inadequate and stunning
the journey is, how it goes both so
slowly and in the blink of an eye—how
dark and then what (against all odds)
so brilliantly lights the way."
>   —Anne Lamott,
>   author of *Grace (Eventually)*

"Dani Shapiro takes readers on an
intense journey in search of meaning

and peace. Her story of hope is eloquently told and unflinchingly honest."

—Jeannette Walls, author of *The Glass Castle*

"This is a book for people who pray, for people who breathe deeply as a form of prayer, for people who have no idea why or even how some people pray. This is a beautiful, wry, and moving story about one intelligent woman's journey into her own life, to the corners where intelligence doesn't always help. *Devotion* is a book for anyone who knows or suspects that they are, to paraphrase Carl Jung, thoroughly unprepared to step into the afternoon of life."

—Amy Bloom, author of *Away*

"I was on the verge of tears more than once in the course of Dani Shapiro's impeccably structured spiritual odyssey. But *Devotion*'s biggest triumph is its voice: funny and unpretentious, concrete and earthy—appealing to skeptics and believers alike. This is a gripping, beautiful story."

—Jennifer Egan, author of *The Keep*

The acclaimed novelist and author of the bestselling memoir *Slow Motion* delivers a searching and timeless new personal account that examines the fundamental questions that keep us awake in the night.

In her mid-forties and settled into the responsibilities and routines of adulthood, Dani Shapiro found herself with more questions than answers. Was this all life was—a hodgepodge of carpools, homework, dinner dates, online shopping, e-mails, meetings, to-do lists? What did it all mean?

Growing up in a deeply religious family that followed traditions instead of subscribing to beliefs, Shapiro had no personal sense of faith, despite repeated attempts to create a sense of connection to something greater. Feeling as if she were plunging headlong into the abyss Carl Jung termed "the afternoon of life," Shapiro wrestled with self-doubt and a searing disquietude that would awaken her in the middle of the night. Scarred by loss—her father's early death, nearly losing her son in infancy, the terminal illness of the mother with whom she shared a troubled relationship—Shapiro had become edgy and uncertain. At the ▶

heart of this anxiety, she realized, was a single question: What did she believe? Seeking to still her apprehension and spurred on by the Big Questions her young son began to raise, Shapiro courageously embarked upon the daunting quest to find meaning in a constantly changing world. The result is *Devotion*: a literary excavation to the spiritual core of a life.

In this voyage of the mind and the heart, Shapiro ponders the varieties of experience that she has pursued—from the rituals of her black hat Orthodox relatives to yoga shalas and meditation retreats to the writings of Abraham Joshua Heschel and Christopher Hitchens. A reckoning of the choices she has made, and the knowledge she has attained, *Devotion* is the story of a woman whose search for meaning ultimately leads her home. A transformative journey at once universal and intensely personal, it offers illumination and inspiration for us all. ∾

Don't miss the next book by your favorite author. Sign up now for AuthorTracker by visiting www.AuthorTracker.com.